MEDICAL
INTELLIGENCE
UNIT

Hematopoietic Stem Cell Development

Isabelle Godin, Ph.D.

Hématopoïèse et Cellules Souches
INSERM U362
Institut Gustave Roussy
Villejuif, France

Ana Cumano, M.D.

Unité du Développement des Lymphocytes
INSERM U668
Institut Pasteur
Paris, France

LANDES BIOSCIENCE / EUREKAH.COM
GEORGETOWN, TEXAS
U.S.A.

KLUWER ACADEMIC / PLENUM PUBLISHERS
NEW YORK, NEW YORK
U.S.A.

HEMATOPOIETIC STEM CELL DEVELOPMENT

Medical Intelligence Unit

Landes Bioscience / Eurekah.com
Kluwer Academic / Plenum Publishers

Printed in the U.S.A.

Kluwer Academic / Plenum Publishers, 233 Spring Street, New York, New York, U.S.A. 10013
http://www.wkap.nl/

Please address all inquiries to the Publishers:
Landes Bioscience / Eurekah.com, 810 South Church Street, Georgetown, Texas, U.S.A. 78626
Phone: 512/ 863 7762; FAX: 512/ 863 0081
http://www.eurekah.com
http://www.landesbioscience.com

Hematopoietic Stem Cell Development, edited by Isabelle Godin and Ana Cumano, Landes / Kluwer dual imprint / Landes series: Medical Intelligence Unit

ISBN: 0-306-47872-2

While the authors, editors and publisher believe that drug selection and dosage and the specifications and usage of equipment and devices, as set forth in this book, are in accord with current recommendations and practice at the time of publication, they make no warranty, expressed or implied, with respect to material described in this book. In view of the ongoing research, equipment development, changes in governmental regulations and the rapid accumulation of information relating to the biomedical sciences, the reader is urged to carefully review and evaluate the information provided herein.

Library of Congress Cataloging-in-Publication Data

Hematopoietic stem cell development / [edited by] Isabelle Godin, Ana
   Cumano.
     p. ; cm. -- (Medical intelligence unit)
   Includes bibliographical references and index.
   ISBN 0-306-47872-2
     1. Hematopoietic stem cells. 2. Hematopoiesis.   I. Godin, Isabelle.
   II. Cumano, Ana. III. Series: Medical intelligence unit (Unnumbered :
   2003)
     [DNLM: 1. Hematopoiesis--physiology. 2. Hematopoietic Stem Cells
   --physiology. 3. Hematopoiesis--genetics.   WH 380 H48692 2006]
   QP92.H4533 2006
   612.4'1--dc22

                                        2005033933

CONTENTS

EDITORS

Isabelle Godin
Hématopoïèse et Cellules Souches
INSERM U362
Institut Gustave Roussy
Villejuif, France
Email: igodin@igr.fr
Chapter 9

Ana Cumano
Unité du Développement des Lymphocytes
INSERM U668
Institut Pasteur
Paris, France
Email: cumano@pasteur.fr
Chapter 9

CONTRIBUTORS

Isabelle André-Schmutz
Laboratoire de Thérapie Cellulaire
  et Génetique
INSERM U 429
Hôpital Necker Enfants Malades
Paris, France
Email:
isabelle.andre-schmutz@nck.ap-hop-paris.fr
Chapter 12

Fumio Arai
Department of Cell Differentiation
The Sakaguchi Laboratory
  of Developmental Biology
School of Medicine
Keio University
Shinjuku-ku, Tokyo, Japan
Chapter 8

Julien Yuan Bertrand
Unité du Développement
  des Lymphocytes
INSERM U668
Institut Pasteur
Paris, France
and
Division of Biological Sciences
University of California, San Diego
La Jolla, California, U.S.A.
Chapter 9

Karine Bollerot
UMR 7622
UPMC Paris VI
Paris, France
Chapter 3

Marina Cavazzana-Calvo
Laboratoire de Thérapie Cellulaire
  et Génetique
INSERM U 429
Hopital Necker Enfants Malades
Paris, France
Email: cavazzan@necker.fr
Chapter 13

Aldo Ciau-Uitz
Institute of Genetics
University of Nottingham
Queen's Medical Centre
Nottingham, U.K.
Chapter 1

Cécile Drevon
Université des Sciences
    et Technologies de Lille
Laboratoire de Biologie
    du Développement
Villeneuve d'Ascq, France
Chapter 3

Elaine Dzierzak
Department of Cell Biology
    and Genetics
Erasmus University Medical Center
Rotterdam, The Netherlands
Email: e.dzierzak@erasmusmc.nl
Chapter 7

Rodolphe Gautier
UPMC Paris VI
Paris, France
Chapter 3

Shubha Govind
Department of Biology
City College of New York
    and Graduate School
University Center of the City University
    of New York
New York, New York, U.S.A.
Chapter 10

Thierry Jaffredo
UMR 7622
UPMC Paris VI
Paris, France
Email: jaffredo@ccr.jussieu.fr
Chapter 3

Olli Lassila
Department of Medical Microbiology
Turku Graduate School
    of Biomedical Sciences
University of Turku
Turku, Finland
Email: olli.lassila@utu.fi
Chapter 4

Jussi Liippo
Department of Medical Microbiology
Turku Graduate School
    of Biomedical Sciences
University of Turku
Turku, Finland
Email: jussi.liippo@utu.fi
Chapter 4

Alexandra Manaia
Unité INSERM U362
Institut Gustave Roussy
Villejuif, France
and
European Learning Laboratory
    for Life Sciences
EMBL
Heidelberg, Germany
Email: manaia@embl.de
Chapter 9

Caroline Marshall
Molecular Immunology Unit
Institute of Child Health
London, U.K.
Email: c.marshall@ich.ucl.ac.uk
Chapter 11

Marie Meister
UPR 9022 du CNRS
IBMC
Strasbourg, France
Chapter 10

Krisztina Minko
Department of Human Morphology
    and Biology
Semmelweis University
Budapest, Hungary
Chapter 3

Katrin Ottersbach
Department of Cell Biology
    and Genetics
Erasmus University Medical Center
Rotterdam, The Netherlands
Chapter 7

James Palis
Department of Pediatrics
Cancer Center
Center of Human Genetics
    and Molecular Pediatric Disease
University of Rochester School
    of Medicine and Dentistry
Rochester, New York, U.S.A.
Email: James_Palis@urmc.rochester.edu
Chapter 5

Roger Patient
The Weatherhall Institute
    of Molecular Medicine
MRC Molecular Haematology Unit
University of Oxford
John Radcliffe Hospital
Headington, Oxford, U.K.
Email: roger.patient@imm.ox.ac.uk
Chapter 1

Stéphane Romero
Laboratoire d'Enzymologie
    et Biochimie Structurale
CNRS
Gif-sur-Yvette, France
Chapter 3

Toshio Suda
Department of Cell Differentiation
The Sakaguchi Laboratory
    of Developmental Biology
School of Medicine
Keio University
Shinjuku-ku, Tokyo, Japan
Email: sudato@sc.itc.keio.ac.jp
Chapter 8

Jeanne Van Celst
Unité INSERM U362
Institut Gustave Roussy
Villejuif, France
Chapter 9

Maggie Walmsley
Institute of Genetics
University of Nottingham
Queen's Medical Centre
Nottingham, U.K.
Chapter 1

Rebecca A. Wingert
Division of Hematology/Oncology
Children's Hospital
Harvard Medical School
Boston, Massachusetts, U.S.A.
Chapter 2

Mervin C. Yoder
Herman B. Wells Center
    for Pediatric Research
Indiana University School of Medicine
Indianapolis, Indiana, U.S.A.
Email: myoder@iupui.edu
Chapter 6

Leonard I. Zon
Division of Hematology/Oncology
Harvard Medical School
and
Howard Hughes Medical Institute
Children's Hospital
Boston, Massachusetts, U.S.A.
Email: zon@enders.tch.harvard.edu
Chapter 2

PREFACE

Adult stem cells are characterized as an autonomous pool capable of regenerating the mature cell compartment of a given tissue, at any time. The hematopoietic system is the best-characterized model in stem cell biology. The progress achieved in the last decades started in the early 1960s with the discovery that multipotentiality and self-renewal were properties of rare individual bone marrow cells. The mechanism of hematopoietic stem cell differentiation and self-renewal, that allow a constant differentiation while maintaining the size of the stem cell compartment, represents still today a major challenge in clinical research. Some of the more recent contributions to this understanding came from the study of the ontogeny of the hematopoietic system.

This book presents the state of the art in developmental hematopoiesis. It includes recent data from various animal models (from Drosophila to higher vertebrates) used to explore different aspects of hematopoietic stem cell (HSC) development. Each model provides information that cannot be easily approached in others, e.g., interspecific chimeras that allow one to follow the fate of engrafted tissue up to adulthood can only be performed in amphibia and avian embryos.

The presented data point to conserved features in developmental hematopoiesis, namely the existence of independent generations of hematopoietic precursors endowed with opposing self-renewal and differentiation properties and with specific roles at different stages of development. In most vertebrates, the first hematopoietic precursors appear in the yolk sac and seem driven to a fast differentiation along the erythromyeloid lineages, at the expense of self-renewal and multipotentiality, while the intraembryonic compartment gives rise to bona fide HSC, precisely in the area that encompasses the aorta, gonads and mesonephros. This scheme constitutes the major advance of the decade in the field of hematology, and has been extended to human embryos.

The experimental systems established to understand the origin of hematopoietic stem cells are now used to approach unresolved questions. While the intraembryonic generation of HSC is well accepted, the nature and differentiation potential of their immediate precursor is still a matter of controversy. The role of environmental signals in the different sites of hematopoietic cell generation has not been analyzed, and the molecular basis of the differences in potential displayed by cells of different origins is, as yet, poorly defined. Another unresolved issue relates to other possible sites that can provide the environmental conditions for HSC generation. The concept that all HSC are generated once in the lifetime, during a short period of development and then exported to hematopoietic organs where they can expand and further differentiate, is consistent with the available data.

However, it is conceivable that other sites throughout embryonic development and even in the adult can provide HSC that contribute to blood cell formation. This volume provides the first elements to answers some of these questions.

The next years will hopefully bring exciting new observations in the field that will allow the more efficient use of these cells in clinical protocols of gene and cell replacement therapy.

Isabelle Godin and Ana Cumano

CHAPTER 1

Ventral and Dorsal Contributions to Hematopoiesis in *Xenopus*

Aldo Ciau-Uitz, Maggie Walmsley and Roger Patient

Abstract

The view that all blood derives from ventral mesoderm has been challenged in recent years. In the *Xenopus* embryo, it is now clear that the embryonic blood compartment, the ventral blood island (VBI), is derived from regions of the pre-gastrula embryo traditionally referred to as dorsal as well as ventral. Furthermore, recent lineage labelling studies in *Xenopus*, show that the adult blood lineage in the dorsal lateral plate (DLP) mesoderm arises independently of the embryonic lineage. Thus, there appear to be three distinct sources of blood in *Xenopus* embryos, two giving rise to the VBI and one the DLP. Distinct origins coupled with separate migration pathways through the embryo suggest that the three populations may be independently programmed during development. Perturbation of BMP signalling shows that all three require this signal in order to form the putative bipotential precursor of blood and endothelium, the hemangioblast. Differences between the embryonic populations and the adult lineage however have been detected with respect to retinoid signalling during gastrulation, and also with respect to specific gene responses to BMP signalling. Experimental manipulations of this model system are beginning to inform our understanding of the developmental programming of hematopoietic stem cells.

Introduction

Blood has been traditionally viewed as a derivative of ventral mesoderm.[1,2] This view in part reflects the location of the blood islands in the *Xenopus* embryo, which are found in the belly region. Early fate mapping of these ventral blood islands (VBI) traced the cells to the opposite side of the pre-gastrula embryo from the Spemann organizer and notochord tissue, traditionally viewed as dorsal.[1] This side of the pre-gastrula embryo was therefore labelled ventral. However, recent studies have uncovered a major contribution to the VBI from the dorsal side of the embryo, rendering the early terminology a little misleading. In addition, the origin of blood cells in the Spemann organizer raises some interesting questions about the embryonic signals required for blood formation, especially with respect to bone morphogenetic protein (BMP) signalling.

Recent data from *Xenopus* have also enlivened the debate about the origin of the adult blood, including the clinically important hematopoietic stem cell (HSC). This has been a controversial issue in mammalian embryos where two models are still competing. The first model was proposed more than 30 years ago and states that the yolk sac is the source of HSCs.[3]

Hematopoietic Stem Cell Development, edited by Isabelle Godin and Ana Cumano.

According to this model, HSCs derive from the yolk sac and enter the blood stream, sequentially seeding the intraembryonic organs where emergence of hematopoietic activity is observed. According to the second model, HSCs first form in the embryo proper independently of the yolk sac.[4] This implies that the precursors of yolk sac and intraembryonic hematopoiesis arise independently during ontogenesis. A resolution of this issue will inform the process of defining the embryonic signals that program the HSC.

Amphibian embryos have been used for a long time as model organisms for developmental biology. Indeed many fundamental concepts, such as the Spemann organizer, were first developed in these organisms. In recent years, the amphibian, *Xenopus laevis*, has dominated studies of gastrulation, the formation of the three primary germ layers, morphogenetic movements and embryonic patterning.

Many characteristics make the *Xenopus* embryo an ideal experimental model.[5] Xenopus embryos can be obtained in large numbers from adult females all year round, and they can be fertilised and cultured in vitro thus providing many synchronously developing embryos. Amphibian eggs are large, being approximately 1 mm in diameter, and therefore easy to manipulate. Because they develop outside the mother and without the need for extraembryonic structures, they are accessible to experimental manipulation at all developmental stages, from oogenesis to adulthood. Because of this and their size, microdissection procedures, such as tissue transplantation, are relatively easy to perform. Development is rapid: the body plan is established and tissue-specific gene activation occurs within 24 h. In addition, the size and number of embryos enable the extraction of sufficient material for biochemical analysis. Finally, the embryos do not grow: their early development consists of a series of cleavage divisions. This means, firstly, that it is possible to construct fate maps by injecting lineage tracers into selected blastomeres: because there is no growth the marker does not become diluted during development.[5] Furthermore, since embryonic blastomeres survive on their yolk reserves, they will divide and even differentiate in a simple buffered salts solution. Thus it is possible to culture explants in isolation and to test defined molecules for their effects on differentiation without the interference of poorly characterized serum components. In fact much of the knowledge of mesoderm induction derives from exposure of naïve ectodermal cells (animal caps) to inducing factor candidates. Another feature that makes *Xenopus* an attractive model for developmental studies is the number of genes characterized, which currently exceeds 5000. Importantly, the genes isolated include factors crucial for hematopoiesis, such as GATA-2, SCL and AML-1/Runx-1 (Xaml).

Hematopoietic Organs in *Xenopus*

During vertebrate development, multiple sites of hematopoiesis are observed. Several tissues within the mammalian embryo serve as reservoirs and/or generators of hematopoietic activity: yolk sac; para-aortic splanchnopleura (P-Sp); aorta, gonads and mesonephros region (AGM); liver; spleen and thymus.[4] In the mouse embryo, the liver, spleen and thymus form relatively late in gestation and are thought to be seeded by hematopoietic progenitors produced elsewhere in the embryo.[6] Shortly after birth, the bone marrow becomes the principal site of hematopoietic activity. The fetal liver is capable of providing all the blood lineages to the developing organism and serves as the source of the HSCs that seed the thymus, spleen and bone marrow.

In avian embryos, the liver is not considered a primary hematopoietic organ. The yolk sac is the organ responsible for erythropoiesis during most of the incubation period. During early stages of development, the yolk sac produces red cells from its own progenitor cells, but for late stages of development, erythropoiesis is carried out by stem cells received from the embryo proper. In the embryo proper, the hematopoietic organs (thymus and bursa of Fabricius) are seeded by HSCs derived from the para-aortic splanchnopleura directly rather than via the liver.[7,8]

Amphibian hematopoiesis is more similar to that of mammalian embryos in that the liver plays a major role. Initially the VBI (yolk sac equivalent) is the major organ for erythropoiesis and also provides hematopoietic precursor cells that seed the liver, thymus and spleen. Erythropoiesis begins in the VBI at stage 28 (31 h) of development and by stage 40 (66 h) the VBI no longer exists.[9,10] In several species of Rana, the second erythropoietic organ is the kidney (pronephros and mesonephros[11]). However, for larval *Xenopus laevis*, the kidney contribution to erythropoiesis represents only 2% of the total number of erythroid cells.[12] The liver initiates erythropoietic activity around stage 46 (106 h), and is considered the major erythropoietic and lymphopoietic organ of the *Xenopus* larva.[12,13] As in the mouse embryo, hematopoiesis in the thymus and spleen is only observed during late stages of development, when the embryo is entering metamorphosis.

In amphibians, as in avians, studies of the development of hematopoietic cells have greatly benefited from the generation of embryo chimeras. Both intra-species and inter-species chimeras can be generated using transplantation techniques. Their use established that two regions of the early *Xenopus* embryo contain hematopoietic precursors:[13] the ventral blood island (VBI), which is located in the ventral mesoderm between the liver anlagen and the proctodeum,[9,11] and the dorsolateral plate (DLP), which is located ventral to the somites and posterior to the pronephros in the lateral mesoderm[14] (see Fig. 2). Subsequently it was demonstrated that the VBI contains the precursors for embryonic and early larval hematopoiesis, whereas the DLP contains the precursors for late larval and adult hematopoiesis.[15-17] Hematopoietic precursors derived from both the VBI and the DLP migrate to and seed the liver, thymus and spleen for a significant period of time.[16] However, the DLP-derived cells are the major contributors to the adult hematopoietic organs, producing blood cells throughout adult life.[14-17] Thus, in terms of their contributions to blood, the *Xenopus* VBI is equivalent to the mammalian yolk sac and the DLP is equivalent to the P-Sp/AGM region.

The relationship between the P-Sp and the AGM in mouse embryos is not yet proven but it is assumed that the P-Sp gives rise to the AGM. A recently described manifestation of the hematopoietic activity in the P-Sp/AGM region is the appearance of clusters in the floor of the dorsal aorta and the other major arteries.[18] Using genes expressed in hematopoietic progenitors as markers, we have recently described intra-aortic hematopoietic clusters in the stage 43 (85 h) *Xenopus* embryo.[19] In addition, others have reported the identification, after stage 44 (90 h) of development, of many leukocytes of the macrophage/monocyte series, located in the mesenchyme underneath the ventral wall of the dorsal aorta.[10] Lineage tracing indicates that the clusters derive from the same blastomere as the DLP (Fig. 2, stages 27-45 and see below).

Origins of Hematopoietic Cells in *Xenopus*

In Xenopus, hematopoietic cells have been traditionally regarded as derived from ventral mesoderm.[1,2] Fate maps established that dorsal and ventral tissues in post-gastrulation embryos derive from opposite sides of the equator (or marginal zone) of the pre-gastrulation embryo.[1,20] Accordingly, the side of the embryo from which dorsal tissues, such as notochord and somites, derived was named the dorsal marginal zone (DMZ) while the side from which ventral tissues, such as blood, derived was named the ventral marginal zone (VMZ). Thus from these fate maps, it was deduced that the VBI (the ventral most mesoderm) derives from the VMZ[1] (Fig. 1A, dark grey blastomere). A small contribution from the DMZ was observed in the experiments of Dale and Slack but has until recently been ignored. It is important also to mention that, for the purposes of this fate map, blood was considered part of lateral plate mesoderm and the origin of the DLP as a separate entity was not addressed.

The view of a ventral origin for all hematopoietic cells in *Xenopus*[21] was first challenged with the discovery that dorsal marginal zone (DMZ) explants cultured in isolation can express

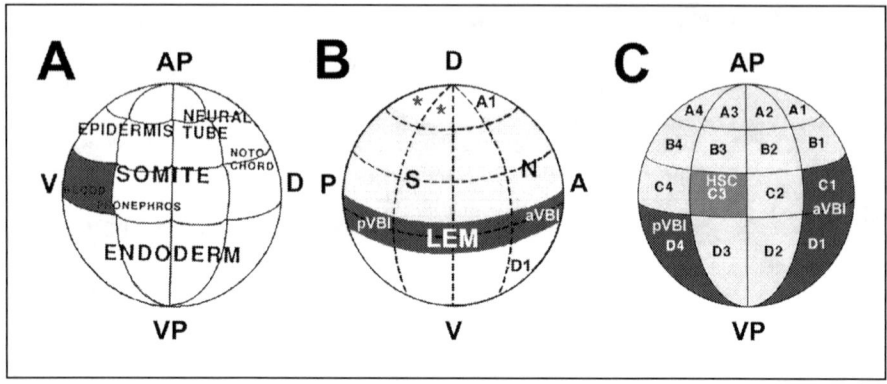

Figure 1. The origins of hematopoietic stem cells (HSC) in *Xenopus*. Evolution of the 32-cell stage fate map for hematopoietic tissues. A) Dale and Slack[1] fate map. Lateral plate, including blood, derives from the most ventral blastomere (dark grey). B) Lane and Smith[23] fate map. The whole of the leading edge of the mesoderm (LEM, darkgrey) contributes to the VBI. Two animal pole blastomeres (stars) also contribute. Note the irregular location of cleavage planes in the vegetal/ventral (V) hemisphere. C) Ciau-Uitz et al[19] fate map. The adult lineage (HSC via DLP and hematopoietic clusters) derives from a blastomere which does not give rise to the embryonic lineage, VBI. The aVBI derives from dorsal whereas the pVBI derives from ventral with no contribution of lateral blastomeres to the VBI. A: anterior; AP: animal pole; aVBI: anterior VBI; D: dorsal; HSC: hematopoietic stem cells; LEM: leading edge of the mesoderm; N: notochord; P: posterior; pVBI: posterior VBI; S: somites; V: ventral; VP: vegetal pole.

Xaml, the Xenopus homolog of the hematopoietic progenitor gene, AML-1 or Runx-1, which is normally expressed in the VBI.[22] Furthermore, LiCl dorsalized embryos were also demonstrated to express both Xaml and α-globin. Interestingly, injection of XRD, a dominant negative form of Xaml, into the VMZ of 4-cell stage embryos, abolished α-globin expression in the most posterior region of the VBI. Conversely, when XRD was injected into the DMZ, α-globin expression was abolished in the most anterior region of the VBI.[22] Taken together, these data suggested that the anterior region of the VBI (aVBI) derives from the dorsal hemisphere of the 4-cell stage embryo whereas the posterior region of the VBI (pVBI) derives from the ventral hemisphere.

These discoveries were confirmed and further expanded by Lane and Smith.[23] These authors demonstrated that C1 (see Fig. 1C for blastomere numbering), the blastomere which gives rise to the Spemann organizer, contributes to primitive blood as indicated by labelled cells in the circulation.[23] Their results also indicated that all the equatorial (tier C) and vegetal (tier D) blastomeres contributed to primitive blood, at least in the vicinity of the cleavage planes between these blastomeres: the region that will constitute the leading edge of the involuting mesoderm (dark grey region in Fig. 1B). They proposed that the most dorsal equatorial blastomeres contribute to the most anterior region of the VBI and that the most ventral equatorial blastomeres contribute to the most posterior region of the VBI, with the equatorial blastomeres in the middle contributing to the central region of the VBI. Thus the concept of a ventral derivation for all hematopoietic cells in Xenopus could only be true if the axes of the 32-cell embryo were rotated through 90° (Fig. 1B). The blood-forming region is represented by the vegetal (yolk-proximal) toroid of cells in the marginal zone (dark grey region in Fig. 1B) with more dorsal mesoderm (somites (S) and notochord (N) in Fig. 1B) derived from the animal (yolk-distal) toroid of cells. The old dorsal side becomes anterior (A) and the old ventral side becomes posterior (P). Such a redrawing of the 32-cell fate map is consistent with the

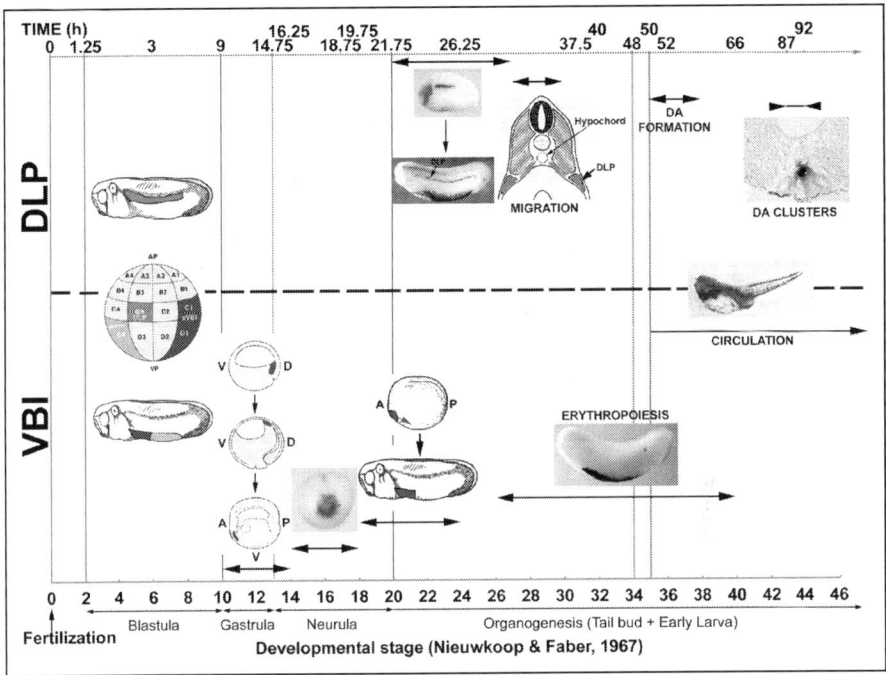

Figure 2. Development of hematopoietic cells at successive sites in *Xenopus*. Precursors to embryonic and adult hematopoietic cells in the 32-cell stage embryo (stage 6). Precursors for the aVBI migrate from dorsal to ventral during gastrulation (stage 10-14). aVBI precursors initiate expression of hematopoietic and endothelial genes in the vicinity of the cement gland, and begin migration into the VBI differentiating into hematopoietic and endothelial cells (stage 14-18). Migrating cells reach the aVBI (stage 24). VBI complete (stage 26). Erythropoiesis (globin synthesis) in the VBI (stage 26-40). Heart starts beating (stage 34, vertical pink line). Blood circulation established (stage 35, vertical red line). Gene expression in the anterior DLP commences (stage 20). DLP complete (stage 26). DLP cells migrate towards the hypochord (stage 27-30). Dorsal aorta (DA) lumen forms (stage 35-38). Hematopoietic clusters associated with the ventral wall of the DA and the underlying mesenchyme form (stage 43-44). A: anterior; AP: animal pole; aVBI: anterior VBI; D: dorsal; DA: dorsal aorta; DLP: dorsal lateral plate; P: posterior; pVBI: posterior VBI; V: ventral; VBI: ventral blood island; VP: vegetal pole.

origins of anterior and posterior somites, with the tail somites coming from the old ventral side and the trunk somites coming from the old dorsal side of the embryo.

Data from our own laboratory confirmed the dual origin of the VBI.[19] Cells derived from the VMZ of the 4-cell embryo populate the pVBI whereas the DMZ forms the aVBI. The progeny of the DMZ occupy the ventral mesoderm from the area of the cement gland to half way down the VBI. At the beginning of gastrulation, the DMZ derivatives include the Spemann organizer, and blastomeres C1 and D1 of the 32-cell stage embryo have been shown to contribute to the Spemann organizer of the gastrulating embryo.[24] Lineage tracing confirmed that the progeny of these blastomeres populate the aVBI of the stage 26 embryo[19] (Fig. 1C). Thus, it appears that the aVBI cells, the "early anterior" Xaml-expressing cells described by Tracey Jr. et al,[22] represent the first fast involuting cells of the gastrula embryo. These involuting cells, initially located dorsally, migrate along the roof of the blastocoel and at the end of gastrulation they are located ventrally, close to the cement gland forming area (Fig. 2, stages 10-14). Later on, they migrate from the cement gland area to their final location, the aVBI (Fig. 2, stages 18-24).

Our 32-cell stage lineage tracing agrees with the Lane and Smith[23] fate map in the sense that a dorsal contribution to the VBI was observed. However, our data do not support their claim that all the blastomeres of the C and D tiers contribute to embryonic blood: we found no contribution from the CD3 blastomere (16-cell stage) or C3 and D3 blastomeres (32-cell stage) to the VBI[19] (Fig. 1C). The differences can be explained by the fact that these authors used irregularly cleaving embryos (compare Fig. 1B with 1C), where blastomeres D2 and D3 are larger than in regularly cleaving embryos, and occupy some of the territory normally contained within D1 and D4 of regularly cleaving embryos.

The VMZ of 4-cell embryos gives rise to both the pVBI and the DLP.[19,25] Thus, at this level of resolution, the adult and one of the two embryonic blood compartments share a common origin. To determine if this resulted from migration between the VBI and the DLP, we performed transplantation experiments.[26] In the case of the DLP, blood cells do not differentiate in situ and its location is defined by tissue transplantation.[14-17] However, the expression of blood genes such as GATA-2 in a strip of mesodermal cells running alongside and immediately dorsal to the developing pronephric duct, and below the somitic mesoderm, enables a visualization of the HSCs in the DLP. With a combination of ventral mesoderm transplantations at the neurula stage (stage 14) of development, and in situ hybridization with a GATA-2 probe at tail bud stages, we were able to demonstrate that no cells migrate from the VBI to the DLP.[26] Thus, the VBI and the DLP derive from different regions of the embryo, at least after stage 14.

To determine the origin of the DLP in embryos earlier than stage 14, and at higher resolution than possible at the 4-cell stage, we fate mapped the progeny of single blastomeres of the 32-cell embryo.[19] To ensure that cells could not arrive via the circulation, embryos were analyzed using blood markers prior to the onset of blood circulation (stage 26). The lineage labeling experiments show quite clearly that the DLP derives from a blastomere (C3) distinct from those (C1, D1, and D4) giving rise to the VBI (Fig. 1C). Thus, at this extremely early stage of development, the precursors to adult and embryonic blood are separate. These results disagree with the fate map of Lane and Smith[23] over the progeny of blastomere C3. In their experiments, C3 gives rise to circulating cells at stage 41, which are known to derive from the VBI,[9,11,16] while our results indicate that C3 gives rise to the DLP and not the VBI. The explanation again appears to reside in the differing cleavage patterns of the embryos injected with lineage label. Lane and Smith never saw regularly cleaving embryos and chose to use embryos with large D2 and D3 blastomeres instead (compare Fig. 1B with 1C). In these irregularly cleaving embryos, blastomeres C3 and D3 occupy territory normally occupied by D4 in regularly cleaving embryos and therefore label the circulating embryonic blood. In contrast, in regularly cleaving embryos, the C3 and D3 blastomeres gave rise to circulating cells at stage 41 in only 2 out of 17 embryos.[19] In conclusion, the results show that precursors for the adult (DLP) and embryonic (VBI) hematopoietic lineages are separate in the early embryo.

In order to follow the fate of the hematopoietic precursors located in the DLP, we monitored C3 injected embryos at increasingly later stages of development.[19] We were able to observe the known migration of cells from the DLP to the midline to later form the dorsal aorta[27] (Fig. 2, stages 27-30). Importantly, the hematopoietic clusters of cells associated with the ventral aspect of the dorsal aorta (Fig. 2, stages 43-44) were also labelled (Fig. 3). Before migration, the cells of the DLP, characteristically for hemangioblasts, coexpress hematopoietic and endothelial genes[19] (unpublished results). However, during the migration phase, expression of hematopoietic genes was silenced but expression of endothelial genes continued. Hematopoietic gene expression returned one day after the aorta had formed. The re-emergence of hematopoietic gene expression was restricted to the single medial dorsal aorta with neither

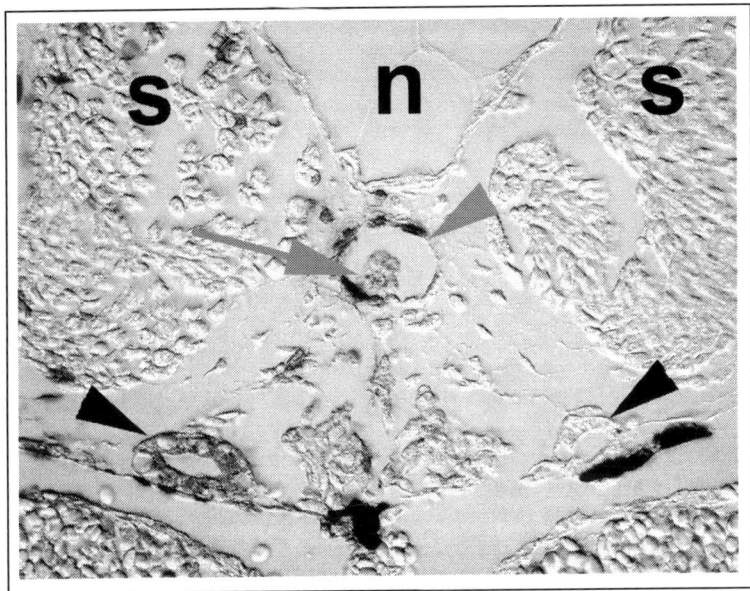

Figure 3. The origins of hematopoietic clusters associated with the ventral wall of the dorsal aorta in *Xenopus*. The hematopoietic clusters and the wall of the dorsal aorta derive from blastomere C3 of the 32-cell stage embryo, the same blastomere that gives rise to the DLP. 10 mm transverse section of a stage 43 embryo injected with β-gal mRNA into one of the two C3 blastomeres at the 32-cell stage. Grey arrow, a hematopoietic cluster. Grey arrowhead, endothelial lining of the dorsal aorta. Black arrowheads, pronephric ducts. n: notochord. s: somite.

the anterior paired dorsal aortae nor the posterior cardinal veins containing hematopoietic clusters. These data probably explain earlier transplantation studies in Rana and *Xenopus* embryos, in which apparently differentiating blood cells derived from the DLP were observed in the vicinity of the dorsal aorta.[14,28,29]

In agreement with our observations in *Xenopus*, evidence that primitive and definitive hematopoietic precursors have different origins, comes from the analysis of interspecific yolk sac chimeras in avian embryos (see chapter by Jaffredo). These experiments demonstrated that adult hematopoiesis originates from intraembryonic precursors and that the yolk sac has an important role only in the establishment of a transient primitive erythropoiesis.

Development of Hematopoietic Cells in Association with Endothelial Cells in *Xenopus*

During ontogeny, the development of the hematopoietic lineage occurs in very close association with the endothelial lineage (see chapter by Jaffredo, Ottersbach and Arai). Within the extraembryonic tissues of mammals and birds, both blood cells and endothelial cells/angioblasts arise within large mesodermal cell aggregates, the blood islands. In the embryo proper, definitive hematopoietic cells also develop in close association with endothelial cells. Definitive hematopoietic cells are thought to first appear as clusters associated with the ventral wall of the dorsal aorta and other arteries of the embryo.[18] The mechanism by which such hematopoietic clusters arise has been a subject of considerable interest and discussion.

Evidence for the existence of a bipotential precursor for hematopoietic and endothelial cells, the hemangioblast, comes from observations that many genes are co-expressed and/or required during early stages of both blood cell and blood vessel development.[6] However, while populations of cells have been shown to give rise to both endothelial and hematopoietic progeny, to date, no in vivo proof of a clonal relationship between the two cell types has been presented. Resolution of the in vivo existence of the hemangioblast will require single cell lineage labelling experiments.

In Xenopus embryos, the two compartments containing hematopoietic precursors, namely the VBI and the DLP, develop in regions where endothelial cells have also been seen developing. For example, the vitelline veins develop near the VBI whereas the posterior cardinal veins develop in the DLP area.[30-32] Our gene expression analysis demonstrates that cell populations co-expressing both hematopoietic and endothelial genes, and, therefore, with characteristics of hemangioblasts, do indeed exist in the early embryo[19] (unpublished results). Although we have not found hemangioblasts in the VBI region itself, a population of hemangioblasts located more anteriorly during neurula stages migrate, as development progresses, and populate the aVBI (Fig. 2, stages 18-24; unpublished results). Lineage labelling experiments show that these cells give rise to both blood and endothelial cells, reinforcing their identity as hemangioblasts.

Regarding the DLP region, a population of cells co-expressing hematopoietic and endothelial genes is also observed there[19] (unpublished results). However, pronephric duct genes are also expressed in these cells at this time[26] (unpublished results). A subgroup of the cells migrates to the midline to give rise, first, to the dorsal aorta and, later, to the hematopoietic clusters associated with the dorsal aorta[19] (Fig. 2, stages 27-30; Fig. 3). Taken together, migration of 'hemangioblasts' appears to be a common feature for both embryonic and adult hematopoietic cell development in *Xenopus*.

Ventral hemangioblasts do not differentiate into blood cells in situ, to do so they migrate posteriorly to the VBI. Similarly, DLP hemangioblasts do not differentiate into blood cells in situ, they migrate to the midline to do so. Interestingly, endothelial cells do differentiate in the places where hemangioblast formation occurs. Cells of the ventral hemangioblast which do not migrate to the VBI, differentiate into angioblasts that later form the ventral aorta and the endocardium. Similarly, cells of the DLP which do not migrate to the midline, differentiate into angioblasts that later form the cardinal veins. Taken together, these observations suggest that the VBI and the dorsal aorta region provide cues that allow the differentiation of hematopoietic cells from hemangioblasts, and that the microenvironments of the locations where the hemangioblasts originate prevent their differentiation into blood.

The posterior component of the VBI (pVBI) derives from ventral regions of the pre-gastrula embryo[19,22,23] (Fig. 1,2). Although hematopoietic and endothelial cells develop in the pVBI region, we have not been able to detect a hemangioblast population which gives rise to this region of the embryo (unpublished results). However, signal perturbations that affect the posterior derivatives affect blood and endothelial cells equally (see below), suggesting that they may well share a bipotential progenitor like the cells in the aVBI and the DLP.

In mammals, the hematopoietic gene, AML-1/Runx1, has a strong association with the adult hematopoietic lineage. Thus, the Runx1 or Cbfb (the heterodimeric partner of Runx1) null mutant mice reveal an absolute requirement for these genes in definitive hematopoiesis, but not in primitive hematopoiesis.[33-36] Runx1 association with the adult lineage is observed at the level of the aortic hematopoietic clusters: Runx1 is strongly expressed by these clusters and an important role there is indicated by their absence in the knockout mouse.[37] In apparent contrast, the Xenopus homologue of Runx1, Xaml, is expressed in the VBI and not in the adult compartment, the DLP.[22] This can be explained by the fact that Runx1 is required for the differentiation of hematopoietic cells.[37,38] Thus, in the VBI, where hematopoietic cells differentiate, Xaml expression is observed and in the DLP, where differentiation of hematopoietic cells does not occur, Xaml is not expressed. However, at later stages of development, Xaml expression is observed in cells of the aortic clusters and the mesenchyme underneath the ventral aspect of the dorsal aorta, confirming its association with differentiating hematopoietic cells and with definitive hematopoiesis.[19] Thus, Runx1's association with the adult hematopoietic lineage extends to amphibians and our data confirm that its expression is coincident with the differentiation of this lineage. Runx1 expression has also been detected in the yolk sac of mouse embryos. Although loss of Runx1 function does not affect the initial wave of erythropoiesis in the yolk sac, it does prevent the later appearance of hematopoietic cells closely associated with the yolk sac endothelium indicating that it may play an important role in endothelial dedifferentiation.[37] Thus, taken together with these data, our observations of the timing of Xaml expression in the dorsal aorta area support the idea that Runx1 facilitates the epithelial to mesenchyme transition that gives rise to hematopoietic cells from the vasculature.[37]

Embryonic Signals Involved in Hematopoietic Development in Xenopus

The distinct origins and migration pathways for embryonic and adult blood suggest that they may come under the influence of different embryonic signals during their ontogeny. Indeed, it has been demonstrated that ventral mesoderm transplanted into the DLP region contributed to adult hematopoiesis and, conversely, that DLP mesoderm transplanted into the VBI region contributed to embryonic blood.[25] These results indicate that hematopoietic precursors located in both the VBI and DLP have the potentiality of producing both embryonic and adult blood and suggest that microenvironmental signals restrict differentiation into one of the lineages.[25]

In Xenopus, mesoderm is induced in the equatorial region of pre-gastrula embryos by TGF-β signals, most likely of the nodal-related family, produced in vegetal cells.[39] Blood is formed from the leading edge of gastrulating mesoderm, the first cells to involute on the dorsal and ventral sides of the embryo[19,23] (Fig. 1B and C). This territory is marked out as early as mid-gastrula by GATA-2, a transcription factor associated with very early blood progenitors including stem cells.[26,40] Blood formation is restricted to this subset of the mesoderm by FGF signalling.[41,42] In isolated VMZ explants grown to tailbud stages, globin expression is restricted to the vegetal end of the explant. When however, VMZs are isolated from embryos injected ventrally with XFD, a dominant negative inhibitor of FGF signalling,[43] globin expression spreads into more animal regions of the explant normally fated to become somites.

Bone morphogenetic proteins (BMPs) have long been associated with ventral patterning and the formation of blood. Thus ectopic expression of BMP4 in *Xenopus* intact embryos, or in animal cap (presumptive ectoderm) explants, stimulates blood cell and globin production,[44-46] and injection of a dominant negative receptor for BMP results in a reduction of blood/globin.[47,48] We have shown that injection of a dominant negative BMP receptor (tBR) into VMZs results in loss of expression of blood and endothelial markers in the posterior blood island compartment whilst leaving the anterior blood island intact, emphasising the distinct origins of the two VBI compartments (unpublished results). Targeting tBR to the DMZ on the other hand, eliminates blood and endothelial marker expression in the anterior blood island and, in addition, reduces expression of blood markers in the posterior compartment suggesting a requirement for a signal from the anterior compartment for posterior VBI development. Furthermore, more lateral injections of tBR into the ventrolateral marginal zone (VLMZ), which includes blastomere C3, eliminated expression of blood and endothelial markers in the DLP, where the precursors for adult blood are located. Thus all three blood compartments require BMP signalling for development of blood.

If the dorsal blood compartment requires BMP signalling, how is this achieved when the precursors of this region originate from the area of the Spemann organizer, a rich source of BMP antagonists such as noggin, chordin and follistatin? Kumano et al[41] demonstrated that even when chordin and noggin were over expressed in organizer blastomeres C1 and D1, these cells were still able to make blood in the anterior blood island suggesting that they were refractory to BMP antagonism. When, however, BMP antagonists were injected into blastomere A4, which is fated to become ventral ectoderm at tailbud stages, globin expression in the VBI was strongly repressed. This suggests that the requirement for BMP in blood formation might occur during gastrulation when the leading edge mesoderm comes into contact with the ventral ectoderm.[41] A role for this ectoderm in BMP signalling to the blood mesoderm has been previously suggested by juxtapositioning animal cap and VMZ explants.[46] Confirmation that the signalling occurs during gastrulation comes from the measurement of the half life of tBR RNA injected into *Xenopus* embryos, which demonstrates that the RNA is abundant at the end of gastrulation but is completely degraded during neurula stages (unpublished results). In addition, any chordin diffusing towards the future ventral ectoderm from the organizer region will be cleaved by the metalloproteases, Xolloid and BMP1, which are expressed there.[49]

The precursors of *Xenopus* adult blood located in the DLP migrate to the midline where they form clusters associated with the dorsal aorta. Whether these clusters emerge from the endothelial cells of the aorta is still debated. The endothelial derivatives of the DLP migrate towards the hypochord, an endoderm derived structure which secretes vascular endothelial growth factor, VEGF.[19,50] The migrating cells express the Flk-1 receptor for VEGF and it is likely that this signalling pathway is influencing their differentiation into endothelial cells.[51] One day later, when VEGF production in the hypochord has ceased, blood gene expression is detected in and under the floor of the dorsal aorta.[19] It is possible that the absence of VEGF and the presence of another signal in the mid-line at this stage of development are factors in the appearance of the clusters. Of particular interest is the expression of high levels of BMP4 in the region just below the dorsal aorta in human embryos (see chapter by Marshall).[52]

Finally, the vitamin A derivative, retinoic acid (RA), has long been known to affect blood cell development. In the *Xenopus* embryo, exposure during gastrulation to exogenous RA results in complete loss of globin expression in the VBI.[26,53] However, GATA-2 expression in both the VBI and the DLP were unaffected, suggesting that elevated RA blocks differentiation of early blood progenitors.[26] We have now re-investigated the action of RA using more blood and endothelial markers to determine which steps in the blood pathway are sensitive to elevated RA (unpublished results). We have shown that all the blood and endothelial markers tested in the DLP, and all except Xaml in the VBI, are expressed normally. We conclude that elevated retinoid signalling prevents blood differentiation from hemangioblasts, possibly by preventing Xaml expression.

To further explore the role of RA signalling in blood development, we treated gastrulating embryos with the drug, Disulphiram, which inhibits aldehyde dehydrogenases, thereby lowering the level of RA signalling (unpublished data). We found that all blood and endothelial gene expression in the VBI was eliminated except for GATA-2. We conclude that formation of embryonic blood and endothelium from GATA-2+ mesoderm requires RA signalling. Interestingly all of the markers were expressed normally in the DLP in 50% of the embryos, while all except XHex and SCL were expressed in the other 50%. We conclude that the DLP is less sensitive than the VBI to reduced RA signalling during gastrulation.

In conclusion, even though the embryonic blood compartment has dorsal as well as ventral contributions, both cell populations are equally sensitive to perturbed BMP or RA signalling. In contrast, the adult blood lineage has a distinct origin to the embryonic blood, and an altered dependence on at least one embryonic signal during ontogeny.

Acknowledgements

The authors would like to thank the MRC, HFSP and CONACyT for financial support.

References

1. Dale L, Slack J. Fate map for the 32-cell stage of Xenopus laevis. Development 1987; 99:527-551.
2. Orkin SH, Zon LI. Genetics of erythropoiesis: Induced mutations in mice and zebrafish. Annual Review of Genetics 1997; 31:33-60.
3. Moore MAS, Metcalf D. Ontogeny of the haematopoietic system; yolk sac origin of in vivo and in vitro colony forming cell in the developing mouse embryo. J Haematol 1970; 18:279-296.
4. Dzierzak E, Medvinsky A, de Bruijn M. Qualitative and quantitative aspects of haemopoietic cell development in the mammalian embryo. Immunol Today 1998; 19:228-236.
5. Smith JC. Mesoderm induction and mesoderm-inducing factors in early amphibian development. Development 1989; 105:665-677.
6. Keller G, Lacaud G, Robertson S. Development of the hematopoietic system in the mouse. Exp Hematol 1999; 27:777-787.
7. LeDouarin NM, Dieterlen-Lievre F, Oliver PD. Ontogeny of primary lymphoid organs and lymphoid stem cells. Am J Anat 1984; 170:261-299.
8. Dieterlen-Lievre F, Godin I, Pardanaud L. Ontogeny of hematopoiesis in the avian embryo: a general paradigm. Curr Top Micro Immunol 1996; 212:119-128.
9. Mangia F, Procicchiani G, Manelli H. On the development of the blood island in Xenopus laevis embryos. Acta Embryologica Experientia 1970:163-184.
10. Ohinata H, Tochinai S, Katagiri C. Ontogeny and tissue distribution of leukocyte-common antigen bearing cells during early development of Xenopus laevis. Development 1989; 107:445-452.
11. Hollyfield JG. The origin of erythroblasts in Rana pipiens tadpoles. Dev Biol 1966; 14:461-480.
12. Ohinata H, Enami T. Contribution of ventral blood island (VBI)-derived cells to postembryonic liver erythropoiesis in Xenopus laevis. Develop Growth Differ 1991; 33(4):299-306.
13. Turpen JB. Induction and early development of the hematopoietic and immune systems in Xenopus. Dev Comp Immunol 1998; 22:265-278.
14. Turpen JB, Knudson CM. Ontogeny of hematopoietic cells in Rana pipiens: Precursor cell migration during embryogenesis. Dev Biol 1982; 89:138-151.
15. Kau C, Turpen JB. Dual contribution of embryonic ventral blood island and dorsal lateral plate mesoderm during ontogeny of hemopoietic cells in Xenopus laevis. J Immunol 1983; 131:2262-2266.
16. Maeno M, Tochinai S, Katagiri C. Differential participation of ventral and dorsolateral mesoderms in the hemopoiesis of Xenopus, as revealed in diploid-triploid or interspecific chimeras. Dev Biol 1985; 110:503-508.
17. Maeno M, Todate A, Katagiri C. The localisation of precursor cells for larval and adult hemopoietic cells of Xenopus laevis in two regions of the embryos. Dev Growth Diff 1985; 27:137-148.
18. Marshall CJ, Thrasher AJ. The embryonic origins of human haematopoiesis. Br J Haematol 2001; 112:838-850.
19. Ciau-Uitz A, Walmsley M, Patient R. Distinct origins of adult and embryonic blood in Xenopus. Cell 2000; 102:787-796.
20. Moody S. Fates of the blastomeres of the 32-cell stage Xenopus embryo. Dev Biol 1987; 122:300-319.
21. Zon LI. Developmental biology of hematopoiesis. Blood 1995; 86:2876-2891.
22. Tracey Jr. WD, Pepling ME, Horb ME et al. A Xenopus homologue of aml-1 reveals unexpected patterning mechanisms leading to the formation of embryonic blood. Development 1998; 125:1371-1380.
23. Lane MC, Smith WC. The origins of primitive blood in Xenopus: implications for axial patterning. Development 1999; 126:423-434.
24. Vodicka MA, Gerhart JC. Blastomere derivation and domains of gene expression in the Spemann Organizer of Xenopus laevis. Development 1995; 121(11):3505-3518.
25. Turpen JB, Kelley CM, Mead PE et al. Bipotential primitive-definitive hematopoietic progenitors in the vertebrate embryo. Immunity 1997; 7:325-334.
26. Bertwistle D, Walmsley ME, Read EM et al. GATA factors and the origins of adult and embryonic blood in Xenopus: responses to retinoic acid. Mech Dev 1996; 57:199-214.
27. Cleaver O, Krieg PA. VEGF mediates angioblast migration during development of the dorsal aorta in Xenopus. Development 1998; 125:3905-3914.

28. Turpen JB, Knudson CM, Hoefen PS. The early ontogeny of hematopoietic cells studied by grafting cytogenetically labeled tissue anlagen: localization of a prospective stem cell compartment. J Dev Biol 1981; 85:99-112.
29. Chen X, Turpen J. Intraembryonic origin of hepatic hematopoiesis in Xenopus laevis. J Immunol 1995; 154:2557-2567.
30. Meyer D, Stiegler P, Hindelang C et al. Whole-mount in-situ hybridisation reveals the expression of the Xl-fli gene in several lineages of migrating cells in Xenopus embryos. Int J Dev Biol 1995; 39(6):909-919.
31. Newman CS, Chia F, Krieg PA. The XHex homeobox gene is expressed during development of the vascular endothelium: overexpression leads to an increase in vascular endothelial cell number. Mech Dev 1997; 66:83-93.
32. Mills KR, Kruep D, Saha MS. Elucidating the origins of the vascular system: A fate map of the vascular endothelial and red blood cell lineage in Xenopus laevis. Dev Biol 1999; 209:352-368.
33. Okuda T, van Deursen J, Hiebert SW et al. AML1, the target of multiple chromosomal translocations in human leukemia, is essential for normal fetal liver hematopoiesis. Cell 1996; 84:321-330.
34. Wang Q, Stacy T, Binder M et al. Disruption of the Cbfa2 gene causes necrosis and hemorrhaging in the central nervous system and blocks definitive hematopoiesis. Proc Natl Acad Sci USA 1996; 93:3444-3449.
35. Wang Q, Stacy T, Miller JD et al. The CBFbeta subunit is essential for CBFalpha2 (AML1) function in vivo. Cell 1996; 87:697-708.
36. Tracey WD, Speck NA. Potential roles for RUNX1 and its orthologs in determining hematopoietic cell fate. Sem Cell Dev Biol 2000; 11:337-342.
37. North T, Gu T-L, Stacy T et al. Cbfa is required for the formation of intra-aortic hematopoietic clusters. Development 1999; 126:2563-2575.
38. Mukouyama Y, Chiba N, Hara T et al. The AML1 transcription factor functions to develop and maintain hematogenic precursor cells in the embryonic aorta-gonad- mesonephros region. Dev Bio 2000; 220(1):27-36.
39. Kimelman D, Griffin K. Vertebrate mesendoderm induction and patterning. Curr Opin Gene Dev 2000; 10:350-356.
40. Walmsley ME, Guille MJ, Bertwistle D et al. Negative control of Xenopus GATA-2 by activin and noggin with eventual expression in precursors of the ventral blood islands. Development 1994; 120:2519-2529.
41. Kumano G, Belluzzi L, Smith WC. Spatial and temporal properties of ventral blood island induction in Xenopus laevis. Development 1999; 126:5327-5337.
42. Kumano G, Smith WC. FGF signaling restricts the primary blood islands to ventral mesoderm. Dev Biol 2000; 228:304-314.
43. Amaya E, Musci TJ, Kirschner MW. Expression of a Dominant Negative Mutant of the FGF Receptor Disrupts Mesoderm Formation in Xenopus Embryos. Cell 1991; 66:257-270.
44. Dale L, Howes G, Price BMJ et al. Bone morphogenic protein-4: a ventralising factor in early Xenopus development. Development 1992; 115:573-585.
45. Jones CM, Lyons KM, Lapan PM et al. DVR-4 (bone morphogenetic protein) as a posterior-ventralizing factor in Xenopus mesoderm Induction. Development 1992; 115:639-647.
46. Maeno M, Ong RC, Xue Y et al Regulation of primary erythropoiesis in the ventral mesoderm of Xenopus gastrula embryo: evidence for the expression of a stimulatory factor(s) in animal pole tissue. Dev Biol 1994; 161:522-529.
47. Graff JM, Thies RS, Song JJ et al. Studies with a Xenopus BMP receptor suggest that ventral mesoderm-inducing signals override dorsal signals in vivo. Cell 1994; 79:169-179.
48. Maeno M, Ong RC, Suzuki A et al. A truncated bone morphogenetic protein-4 receptor alters the fate of ventral mesoderm to dorsal mesoderm—roles of animal pole tissue in the development of ventral mesoderm. Proc Natl Acad Sci USA 1994; 91(22):10260-10264.
49. Goodman SA, Albano R, Wardle FC et al. BMP1-related metalloproteinases promote the development of ventral mesoderm in early Xenopus embryos. Developmental Biology 1998; 195(2):144-157.
50. Cleaver O, Tonissen F, Saha MS et al. Neovascularization of the Xenopus embryo. Dev Dyn 1997; 210:66-77.

51. Eichmann A, Corbel C, Nataf V et al. Ligand-dependent development of the endothelial and hemopoietic lineages from embryonic mesodermal cells expressing vascular endothelial growth factor receptor 2. Proc Natl Acad Sci USA 1997; 94:5141-5146.

52. Marshall CJ, Kinnon C, Thrasher AJ. Polarized expression of bone morphogenetic protein-4 in the human aorta-gonad-mesonephros region. Blood 2000; 96(4):1591-1593.

53. Sive HL, Cheng PF. Retinoic acid perturbs the expression of Xhox.lab genes and alters mesodermal determination in Xenopus laevis. Genes Dev 1991; 5:1321-1332.

CHAPTER 2

Genetic Dissection of Hematopoiesis Using the Zebrafish

Rebecca A. Wingert and Leonard I. Zon

Abstract

Hematopoiesis during embryogenesis and adult life has been extensively studied in several vertebrate models. The zebrafish (Danio rerio) has recently provided a powerful genetic system to further dissect the molecular pathways of hematopoiesis. Large-scale genetic screens utilizing the zebrafish have identified blood mutants with defects in mesoderm patterning, hematopoietic stem cell generation and maintenance, hematopoietic progenitor cell proliferation and differentiation, and erythoid differentiation. Cloning of the mutations has revealed both known and novel genes required for blood production, and served to underscore the strength of the zebrafish as a hematopoietic model. Continued study of zebrafish hematopoietic mutants offers an exciting venue to investigate hematopoiesis during vertebrate ontogeny, as well as to elucidate therapeutics for human blood diseases.

Introduction

Hematopoiesis is a dynamic, ongoing process that generates all blood cell lineages from the pluripotent hematopoietic stem cell (HSC) throughout the life of an organism. The zebrafish (Danio rerio) has emerged as a powerful model system to study vertebrate organ development and heritable human disorders.[1-4] In particular, zebrafish are useful to study the development of the hematopoietic system.[5-6] Zebrafish genetic screens have found mutations that affect a broad range of events involved in hematopoiesis. Progressive improvements in the zebrafish genomics infrastructure, including the creation of physical and genetic maps, radiation hybrid panels, and large genomic insert libraries, have enabled the genetic cloning of many zebrafish mutants.[7-14] A number of zebrafish hematopoietic mutants have been identified by a combination of candidate and positional cloning strategies. Characterization of these blood mutants has revealed new perspectives on hematopoiesis, and in several cases, mutants represent new models of human blood diseases. This review will address the zebrafish blood mutants that have been generated to date and the insight that their embryological and genetic investigations have provided into hematopoiesis.

Zebrafish Hematopoiesis

As in other vertebrates, hematopoiesis in zebrafish ontogeny occurs in multiple waves from different anatomical sites in the embryo.[15,16] An initial primitive wave, equivalent to mammalian yolk sac hematopoiesis, occurs within the zebrafish embryo proper. Mesoderm

Hematopoietic Stem Cell Development, edited by Isabelle Godin and Ana Cumano.
©2006 Eurekah.com and Kluwer Academic / Plenum Publishers.

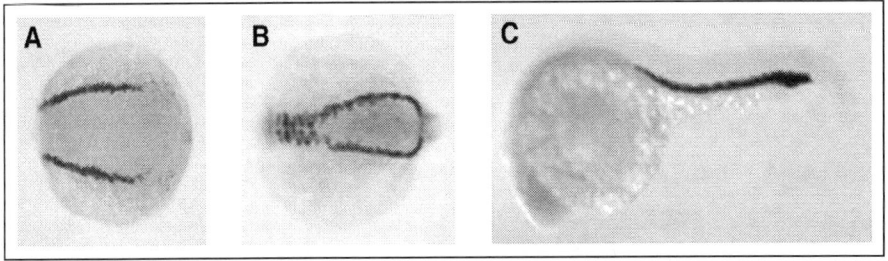

Figure 1. Formation of the intermediate cell mass. Following mesoderm induction, the ventral mesoderm domain will give rise to the primitive blood. In situ hybridization for *scl* expression at 12 hpf marks the stripes of ventral mesoderm blood precursors in the zebrafish embryo (A). B) *GATA-1* expression in the ventral mesoderm at 16 hpf as the stripes fuse, and (C) at 24 hpf after the ICM has formed (and before erythrocytes have exited into circulation).

induction and subsequent patterning divides the mesoderm into dorsal and ventral domains. Following gastrulation, the ventral mesoderm that will give rise to the blood consists of two stripes of cells that migrate medially to form the intermediate cell mass (ICM), the intraembryonic blood island which contains both hematopoietic and vascular progenitors (Fig. 1A-C).[15,17,18] Primitive erythroid cells exit the ICM and migrate rostrally along the trunk to enter circulation in a path over the yolk sac concomitant with the onset of rhythmic heart contractions at 24 hours post fertilization (hpf).[17] Embryonic circulation is comprised of erythroid cells, macrophages, and neutrophils.[19-21] Like mammals, the primitive erythroid cells express embryonic globins.[22] As development progresses, there is a transition to definitive hematopoiesis which begins as early as 4 days post fertilization (dpf) in the kidney primordium.[18] Definitive zebrafish hematopoiesis produces cells of the erythroid, myeloid, lymphoid, and thrombocyte lineages; these cells arise primarily from the adult kidney, though some hematopoietic activity occurs in the adult spleen (Fig. 2D).[15,18] Zebrafish peripheral blood cells resemble mammalian cells, but definitive erythrocytes retain their nuclei as in other fish.[23,24] Erythropoiesis in adult zebrafish is characterized by the expression of adult globins, demonstrating that developmental globin switching occurs in the zebrafish.[25]

Zebrafish orthologues of genes required for both primitive and definitive hematopoiesis, including scl, lmo2, GATA factors, c-myb, runx1, pu.1, ikaros, rag, and globins, have been isolated and function similar to their mammalian counterparts.[17,19,20,22,25-33] The expression of these genes has been used to follow the developing blood populations during the primitive and definitive waves by whole-mount RNA in situ hybridization. The expression of scl begins at 2-3 somites in the stripes of ventral mesoderm and continues to be expressed in the blood until 24 hpf.[26,28] Other markers of early hematopoietic progenitors, including lmo2, GATA-2, and hhex are expressed in the ventral mesoderm and in the ICM compartment with the same timing as scl expression.[17,31,34] A subset of the scl-expressing cells begin to express GATA-1 at the 4-5 somite stage; coexpression of scl and GATA-1 delineates the primitive erythroid population within the ICM, while the gata1-negative cells in the ICM represent angioblasts.[17,28] Markers of the differentiated erythroid lineage, including the embryonic globin chains, heme biosynthetic enzymes, and erythroid cytoskeletal genes, are expressed in the ICM at 12 to 16 somite stages.[22,29,35] The definitive wave is marked by expression of c-myb between approximately 31 and 38 hours in the ventral wall of the dorsal aorta, the zebrafish equivalent to the mammalian aorta-gonad-mesonephros (AGM) region (Fig. 2A-B).[18,31] By 4 dpf, hematopoietic stem cells (HSCs) have colonized the developing kidney and have begun to produce a new population of circulating erythrocytes (Fig. 2C).[15,18] Also beginning at 4 dpf, morphologically distinct definitive erythroid cells are observed in circulation; these definitive

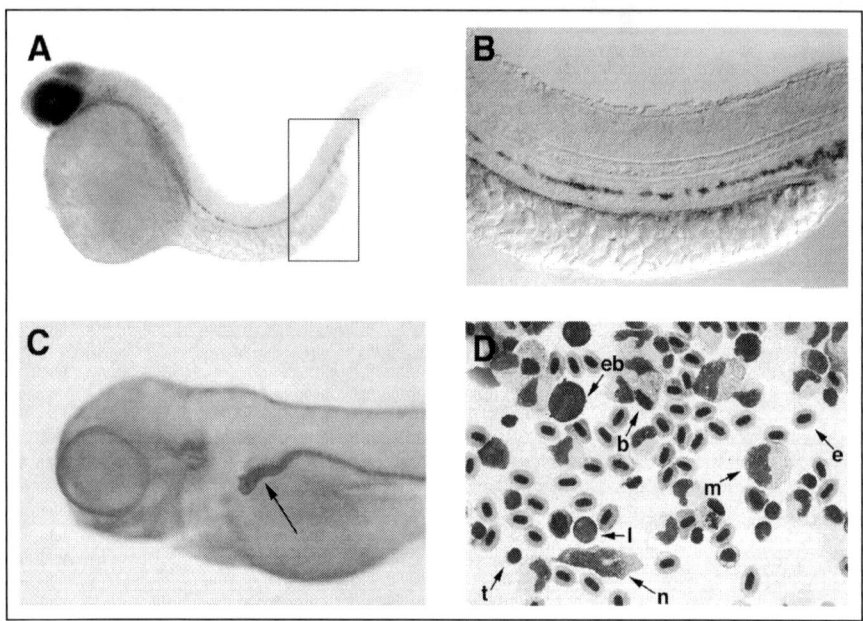

Figure 2. The site of hematopoiesis changes during zebrafish ontogeny. A) At 36 hpf, expression of *c-myb* in the wall of the dorsal aorta marks the zebrafish equivalent to the mammalian AGM, enlarged in (B). C) Arrow indicates the location of the embryonic kidney at 5 dpf, marked by expression of Na$^+$/K$^+$ ATPase α1 subunit. D) Whole kidney marrow from a zebrafish adult contains eo/basophils (b), erthroblasts (eb), erythrocytes (e), lymphocytes (l), monocytes (m), and thrombocytes (t).

cells likely arise from the AGM, though lineage tracing has not been done.[36] This second population of erythrocytes still express embryonic globins; adult globin expression is detectable at approximately 25 dpf, suggesting that zebrafish actually undergo an embryonic to larval to adult globin transition (A. Brownlie and L.I. Zon, unpublished). The thymic organs form by 65 hpf and are then populated by rag1, lck, and ikaros-expressing lymphocytes at 4 dpf.[18,32,33,37,38] Macrophages arise independently from a rostral anlage that is derived from anterior paraxial mesoderm.[19,20] The macrophage anlage is marked by expression of pu.1, fms, leukocyte-specific plastin (l-plastin), c-myb, and draculin, with expression of pu.1, l-plastin, and fms distinguishing myeloid cells from the primitive erythroid population.[19-21,30,39,40] Macrophages are present on the yolk sac as early as 24 hpf, though it is unknown whether they represent a primitive macrophage population that is later replaced by monocyte-derived macrophages or whether they remain as long-term residents in the developing fish.[19]

Zebrafish Genetics and Hematopoietic Mutants

The zebrafish has several attributes that make it an excellent vertebrate system to study the developmental biology and genetics of hematopoiesis. Since zebrafish fertilization is external, embryos are accessible at all stages of development. The transparency of zebrafish embryos permits visual observation of circulating blood cell morphology (number and color) from the time circulation begins at 24 hpf up to day 7-10 of development. Blood cells can be obtained easily from the zebrafish embryo for analysis, and as discussed, zebrafish hematopoiesis parallels that of higher vertebrates. Zebrafish are amenable to forward genetic screens as a large number of adults can be maintained in a relatively small area, adults generate hundreds of offspring in a single mating, and the fish have a short generation time to sexual maturity (approximately 3 months). Additionally, mutations in the zebrafish germ line can be induced by chemical mutagens, gamma or X-rays, and more recently via insertional mutagenesis.[41-46]

Two large-scale genetic screens have used zebrafish to identify genes required for development.[47,48] From these screens, over 50 blood mutants comprising 25 complementation groups were identified by inspecting zebrafish embryos for the absence of circulating blood or an alteration in its color.[36,49] In addition, several spontaneous zebrafish blood mutants have been identified, as well as blood mutants generated serendipitously in other screens.[50-52] This combination of sources has produced a plethora of hematopoietic mutants that can be placed into the following categories: mesoderm patterning mutants, hematopoietic stem cell (HSC) mutants, committed progenitor mutants, proliferation and/or maintenance mutants, hypochromic mutants, and photosensitive mutants (Fig. 3). The embryological characterization and genetic cloning of these blood mutants by several groups has revealed valuable perspective into the molecular regulation of hematopoiesis.

Mesoderm Patterning Mutants

Mesoderm induction and patterning during embryogenesis establishes early dorsal and ventral mesoderm domains in the blastula; blood is specified subsequently in the ventral domain.[53] Signaling through members of the transforming growth factor β-related bone morphogenic protein (BMP) family are necessary for correct specification of ventral mesoderm.[54,55] Secreted BMP-2 and BMP-4 signal through BMP receptors, leading to the downstream activation of SMAD proteins.[56] SMAD proteins function in turn to activate transcription of target genes, which include Mix.1 and its family members. Negative regulation of BMP signaling occurs via the action of competitive antagonists for the BMP receptors. These antagonists are secreted from the Spemann organizer to induce specification of dorsal mesoderm identity, and include the genes chordin, noggin, cerberus, and follistatin.[57]

Zebrafish with mutations that alter BMP signaling have defects in dorsoventral pattern formation that result in the inability to induce normal hematopoiesis in the ventral mesoderm. The mutants swirl (swr) and snailhouse (snh) represent respective mutations in the ligands BMP2b and BMP7, and have expanded dorsal tissues at the expense of the ventral blood and pronephros.[58-61] The dorsalized mutant somitabun (sbn) also has greatly reduced blood.[58] sbn is caused by a defect in the transcription factor smad5, which is specifically required downstream of Bmp2b signaling to mediate Bmb2b autoregulation.[62] The dorsalized lost-a-fin (laf) mutant represents a mutation in activin receptor-like kinase 8 (alk8), a novel type 1 serine/threonine BMP receptor.[58,63,64] Embryos lacking both maternal and zygotic laf expression mimic the strongly dorsalized phenotype of swr and snh mutants, evincing a complete absence of blood.[64] In contrast, dino mutants, which have a mutation in the BMP antagonist chordin, display a strong ventralized phenotype with expansion of blood in the embryo.[65,66] Phenotypic analysis of these zebrafish mutants supports the current model of how BMP signaling acts to direct mesoderm dorsoventral fate. However, the minifin (mfn) mutant, which has a defect in the tolloid gene, has normal primitive blood production even though embryos are dorsalized and suffer a variable loss of tail tissue.[58,67] Since overexpression of zebrafish tolloid causes a moderate ventralization of the embryo along with expansion of the blood, the normal blood in mfn mutants might be explained by the presence of another tolloid (or tolloid-like) gene.[68] The remaining dorsalized ogon (ogo) and ventralized piggytail (pgy) mutants are likely to harbor mutations in genes that affect BMP signaling in the early embryo; recent evidence suggests that the ogon gene in fact acts to antagonize BMP signaling through a tolloid- and chordin-independent function.[58,65,69,70]

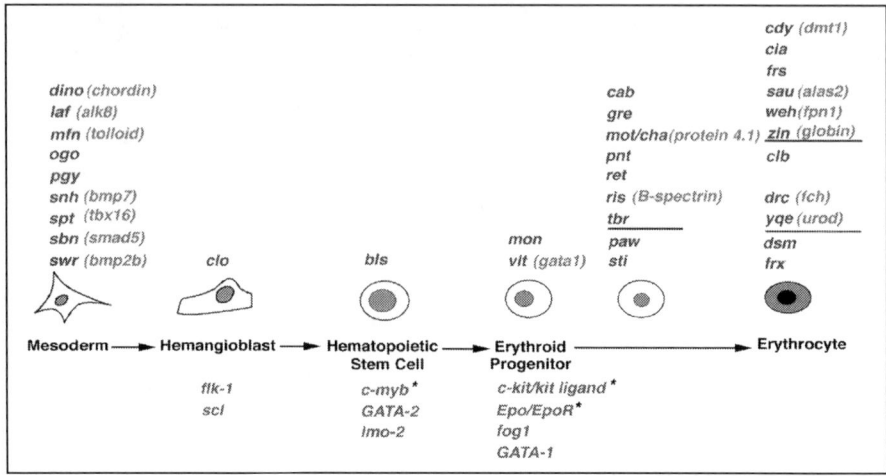

Figure 3. Zebrafish blood mutants represent defects at various stages of primitive hematopoiesis. Hematopoietic mutants identified in two large-scale screens for genes affecting embryogenesis have been classified based on the nature of their primitive blood defects into the following categories: mesoderm patterning mutants, hematopoietic stem cell mutants (inclusive of hemangioblast defects), erythroid progenitor mutants, and erythroid differentiation/proliferation mutants (inclusive of hypochromic and photosensitive mutant classes). Complementation analysis has not been performed between the *paw, sti, clb, dsm,* and *frx* (isolated in Boston) and mutants of each respective phenotype category, so these mutants may be in fact allelic to other members of their category. Top panel: black: zebrafish mutant; grey between bracket: identified gene; Bottom panel: genes required for primitive hematopoiesis; *: genes required at the same stages for definitive hematopoiesis.

In addition to BMP signaling, the T-box transcription factor VegT (tbx16/Brat) is crucial for endoderm formation and the release of mesoderm-inducing signals. The zebrafish spadetail gene appears to represent a homologue of VegT, and consistent with this, spt mutants develop with severe deficiencies in both mesodermal and endodermal tissues.[50,72,73] spt embryos have defective convergence of their trunk paraxial mesoderm; the mesoderm defects in spt also result in poorly formed trunk somites and disorganized tail development, the latter including defects in both blood and vessel formation.[31,50,72] spt embryos have reduced expression of lmo2 and GATA-2 and in the ventral mesoderm, and the stripes of blood progenitors which are present fail to fully converge at the midline (Fig. 4).[31] Few GATA-1-expressing cells are seen in the ICM of spt embryos, and embryos make hardly any circulating primitive erythroid cells.[31] spt embryos do express blood vessel markers, including fli1, flk1, and flt4, though expression (as with eventual vessel structure) is disorganized in the embryo tail. While it is unclear precisely how the spt gene functions to specify blood fate, the blood phenotypes observed in both spt and the various dorsoventral mutants underscore the importance of proper mesoderm patterning for normal embryonic hematopoiesis.

Hematopoietic Stem Cell Mutants

A number of zebrafish mutants have been collected that correctly pattern their mesoderm, but which possess defects in the development or maintenance of hematopoietic stem cells (HSCs). During ontogeny, HSCs have been hypothesized to originate from a precursor cell, coined the hemangioblast, that is capable of producing both the blood and vessel endothelial lineages.[74] A close anatomical association exists between blood and vascular cells during embryogenesis.[74-78] Several lines of experimental evidence support the existence of a bipotential hematopoietic-endothelial precursor, which functionally explains the intimate developmental proximity of these tissues.[79] Blood and endothelial progenitors have been found to coexpress a number of genes, including CD34, c-kit, flk-1, scl, tie-1, and tie-2.[80-88] Mice lacking the flk1, CBF, or scl genes have both blood and vascular defects.[89-91] Mouse chimeric transplantation experiments have found that flk1 deficient embryonic stem cells fail to contribute to blood and vascular lineages.[85] Finally, in vitro culture of blast-colony-forming cells in the presence of appropriate cytokines has found that these cells have the capacity to give rise to both hematopoietic and endothelial cells.[92-94]

The zebrafish mutant cloche (clo) has defects in its hematopoietic and endothelial lineages during embryogenesis, suggesting that clo is essential for the formation or maintenance of the hemangioblast. clo embryos almost entirely lack blood and fail to develop vasculature; in addition, the heart chambers become enlarged due to the lack of an endocardium, with the atrium expanded to resemble a bell-like shape.[51] clo embryos have severely reduced scl, lmo2, GATA-2, and GATA-1 expressions, suggesting a loss of blood progenitors (Fig. 4).[26,28,31,51,95] In addition, clo embryos have almost undetectable expression of flk1, and tie-1 expression is absent, consistent with the loss of vessels.[31,95] Overexpression of the bHLH transcription factor scl in clo mutants has been shown to rescue blood expression of GATA-1 and vessel expression of both flk1 and tie1, though scl has been eliminated as a genetic candidate for clo based on the respective linkage group positions of scl and clo.[28] The rescue of blood and vessel structures by scl overexpression in clo mutants establishes that clo acts prior to scl to specify the fate of these tissues. Remarkably, SCL variants unable to bind DNA rescue primitive erythropoiesis and vasculogenesis in clo embryos, demonstrating that scl functions in a DNA-binding-independent manner to enable development of blood and vessel lineages.[96] hhex overexpression in clo mutants has been shown to rescue expression of scl, GATA-1, flk1, flil, and tie1.[34] However, hhex has been elimininated as a candidate for clo based on their respective genomic loci.[34] Furthermore, despite the striking similarity between the clo phenotype and the mouse flk-1 knockout phenotype, flk-1 also maps independently of clo. clo is likely to function prior to flk1 to specify vascular differentiation, as flk1 expression is not detected in clo mutants where endothelial cells and their progenitors normally reside.[95] Overexpression of bmp4 in clo embryos fails to rescue scl or GATA-1 expressing cells, indicating that the clo gene product is required for hematopoietic patterning subsequent to ventral mesoderm induction.[52] Thus, clo gene function has been positioned after ventral mesoderm patterning but before the functions of scl and flk1 to direct blood and vessel differentiation. Additionally, the clo gene product is required both cell-autonomously and non-cell autonomously for blood differentiation and survival.[97] Unfortunately, the isolation of the clo gene has proven difficult due to its telomeric location; the successful cloning of this gene would clearly provide invaluable insight into the early mechanisms regulating embryonic development and differentiation of the hematopoietic and endothelial lineages.

The bloodless (bls) mutation (also known as sort-of-bloodless) causes an absence of primitive erythrocytes in a dominant but incompletely penetrant manner (Fig. 4).[52] Severely affected bls embryos have virtually no erythroid cells at the onset of circulation; however, bls mutants begin to recover circulating red cells after 5 dpf, coinciding with the initiation of definitive blood cell production around 4 dpf, and mutants live readily to adulthood. Primitive macrophages develop normally in bls mutants, and lymphopoiesis is normal, albeit delayed: lymphoid precursors are absent in the thymic epithelium at 4.5 dpf, but lymphocytes populate the thymic epithelium by 7.5 dpf. bls mutants have greatly reduced scl expression throughout embryogenesis, suggesting that the bls gene product is required for hematopoietic stem/progenitor cell specification. Up to 18 hpf, GATA-1 transcripts are also greatly reduced in bls mutants, after which GATA-1 ICM expression is undetectable by in situ hybridization.

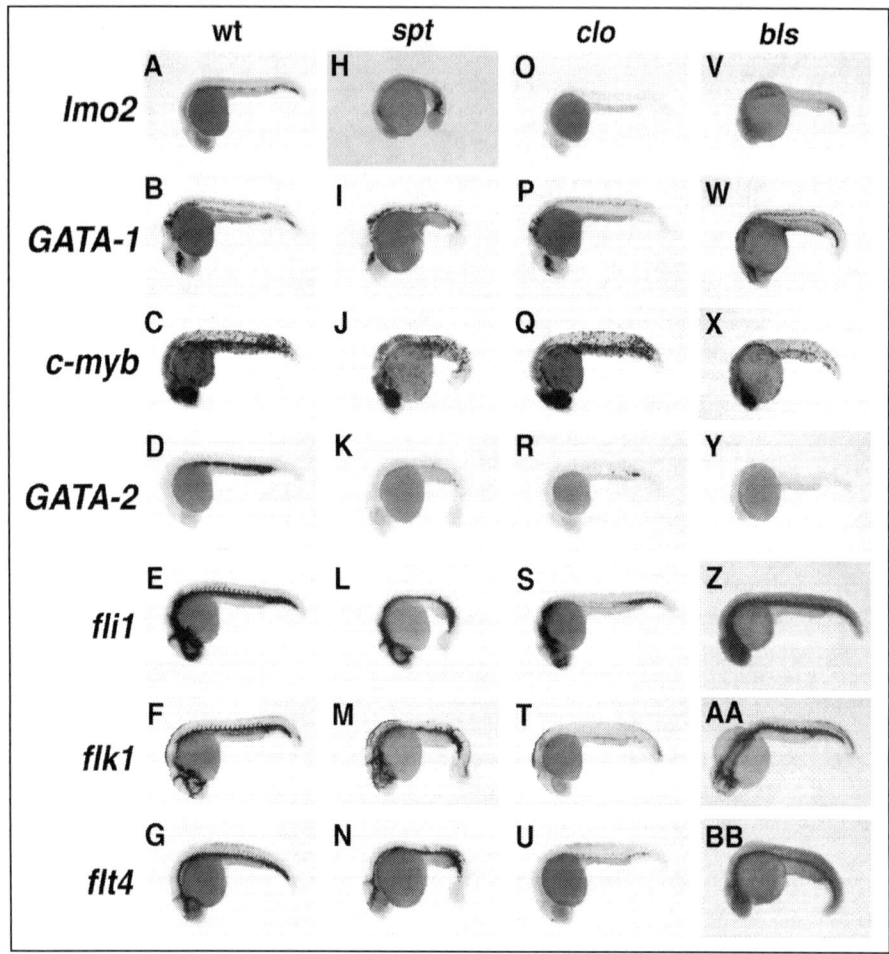

Figure 4. Characterization of hematopoiesis in zebrafish embryos at 24 hours post fertilization. The expression of *lmo2* and *GATA-2* (A, D) delineate the blood progenitors in the ICM, while expression of *GATA-1* and *c-myb* (B, C) mark the primitive erythroid cells . Vascular markers include *fli1*, *flk1*, and *flt4* (E-G). *clo* embryos lack blood and blood vessels, and accordingly lack expression of early hematopoietic and vascular markers (O-U). *spt* and *bls* embryos have angioblasts (though in *spt* they are disorganized) (L-N, X-BB), and express *lmo2* and *GATA-2*, but lack *GATA-1* expression.

Interestingly, bls mutants express GATA-2 in their ICM, though the level is reduced in comparison to wild-type siblings. The GATA-2-expressing cells could largely represent endothelial cells in the ICM region, since GATA-2 is expressed in both blood and endothelial progenitors, and the ICM is known to contain both populations.[78] In agreement with this, bls mutants have morphologically intact vasculature, and maintain normal flk1 expression during embryogenesis. The few scl-expressing cells that are specified from the ventral mesoderm and then fail to maintain GATA-1 eventually undergo apoptosis in the ICM between 15 and 23 hpf. The overexpression of scl (but not bmp4 or GATA-1) in bls mutants can rescue embryonic erythrocyte production in the ICM; scl overexpression could act to rescue hematopoietic progenitors or potentially expand the number of blood progenitors specified from the mesoderm.

Notably, scl overexpression in clo mutants appears to rescue GATA-1 expression in more ICM cells than scl overexpression does in bls mutants, implying that the bls gene may play a role in maintaining scl expression in blood progenitors. The inability of GATA-1 overexpression to rescue embryonic erythrocytes suggests that additional downstream targets of both scl and bls are required for primitive erythroid hematopoiesis. Cell transplantion experiments found that bls embryo donor cells contributed to the GATA-1-expressing ICM population in wild-type hosts, whereas wild-type donor cells were unable to produce GATA-1-positive erythroid progenitors in bls mutants. The non-cell autonomous requirement for the bls gene combined with the incomplete penetrance of the bls phenotype supports the speculation that bls is a secreted factor or cell surface receptor that is uniquely necessary for development or mainte-nance of primitive hematopoietic precursors, possibly by regulating scl expression. Normal definitive hematopoiesis in bls fish indicates that HSCs are correctly developed in spite of the bls mutation, and emphasizes that bls is strictly required for the generation or maintenance of hematopoietic precursors during the primitive wave.

The clo and bls mutants demonstrate the utility of zebrafish blood mutants to study the development and regulation of hematopoietic stem cells in the early vertebrate embryo. The eventual cloning of the clo and bls genes will allow further analysis of their respective roles in maintaining primitive hematopoiesis. At least two additional mutants have been described as having a bloodless phenotype at 24 hpf.[36] The early bloodless phenotype of these mutants could indicate defects at the HSC or hematopoietic progenitor cell level. Based on murine knock-out models, an scl, lmo2, or GATA-2 genetic mutant should fail to make HSCs properly, and would be mostly likely bloodless at 24 hpf. The analysis of gene expression in the remaining bloodless mutants will allow their defects to be specifically categorized.

Committed Progenitor Mutants

The vlad tepes (vlt) and moonshine (mon) mutants also fail to properly generate appreciable circulating blood cells at the onset of circulation, and remain largely anemic during development.[36,49] Based on the expression of early blood markers, both mutants appear to make hematopoietic progenitors, but the progenitors are unable to give rise to the wave of primitive erythrocytes. vlt has been shown to represent a nonsense mutation in GATA-1 that results in a truncated protein which is unable to bind DNA or mediate GATA-specific transactivation.[98] Expression of scl, lmo2, and GATA-2 are normal in vlt embryos until 24 hpf, indicating the presence of hematopoietic progenitors in the ICM. vlt embryos proceed to initiate normal levels of ICM GATA-1 expression, but later lose GATA-1 expression by 26 hpf; embryonic α-globin gene expressions are retained in the ICM, though other markers of dif-ferentiated erythrocytes, including band3 and alas2, are markedly absent in vlt mutants. The myeloid markers pu.1, l-plastin, and c/ebp1, and lymphoid markers ikaros and rag1 are all expressed normally, evidence that these lineages are unaffected by the vlt mutation. The loss of GATA-1 function is consistent with a specific loss of the primitive erythroid lineage, as it recapitulates the red cell precursor arrest observed in murine studies when GATA-1 function has been removed.[99-102] An upregulation of GATA-2 expression has been proposed to compensate for null GATA-1 function in the murine knockout; increased GATA-2 expression is not observed in vlt mutants, though predicted gata targets (such as α-globins) are still expressed at relatively high levels in vlt embryos. The embryonic lethality of vlt embryos mirrors the lethality of GATA-1-/- mice.[99]

The mon mutant (also isolated as the mutant vampire) is bloodless at the onset of circulation.[36,49] Like the vlt mutant, mon specifically affects the development of the erythroid lineage during hematopoiesis. mon embryos have normal ICM expression of GATA-2 until 18 hpf. GATA-1 expression is strongly reduced in weaker mon alleles and absent in strong mon alleles, while markers of all other blood lineages appear normal (D. Ransom and L.I.

Zon, unpublished).[49] In addition to affecting the formation of erythroid cells, mon embryos have abnormal migration of neural-crest derived iridophores while melanophores are normal.[103] mon embryos also suffer from fin necrosis, resulting in the appearance of jagged looking fins.[49,103] The combined blood and neural crest defects in mon embryos are reminiscent of the phenotypes observed in the murine steel and white-spotting mutants that have respective defects in c-kit and its receptor; the c-kit mutation in zebrafish, however, causes defects in migration and survival of neural crest melanocytes but does not display a hematopoietic defect.[104] The correlation between mon allele strength and GATA-1 expression could indicate that mon plays a role in regulating GATA-1 expression. The mon and vlt mutants serve to demonstrate that an early bloodless phenotype is equally likely to represent a primitive erythroid lineage defect as a HSC defect in the embryo.

Differentiation and Proliferation Mutants

A large category of blood mutants have a decreased blood phenotype as embryonic development proceeds. These mutants are characterized by normal initiation of hematopoiesis that is followed by a dramatic decrease in the number of circulating blood cells after 3 days. As such, the expression of early blood markers, including GATA-1, is normal, and wild-type numbers of circulating erythroid cells are seen at 24 hpf. The subsequent decrease in circulating cells in these mutants suggests a defect in erythroid survival or in hematopoietic progenitor proliferation after 24 hpf. The class includes the mutants cabernet (cab), chablis (cha), grenache (gre), merlot (mot), pale and wan (paw), pinotage (pnt), retsina (ret), riesling (ris), sticky blood (sti), and thunderbird (tbr).[36,49] The class can be subdivided into mutants with an onset of anemia at 2 dpf (gre, pnt, sti, tbr) and those that become anemic between 3-5 dpf (cab, cha, mot, paw, ret, ris).[36,49] The cloning of two late decreasing mutants, ris and mot/cha (since discovered to be allelic), have found defects in erythroid cytoskeletal proteins that are necessary for maintaining the integrity of the red cell membrane and hence red cell survival.[29,105]

ris represents a null mutation in erythroid β-spectrin.[29] Erythroid spectrins are the largest proteins of the red cell cytoskeleton and are responsible for providing the framework and stability of the cell membrane.[106] Individual α and β-spectrins form antiparallel dimers which self associate to form $\alpha_2\beta_2$ tetramers; β-spectrin binds to ankyrin and protein 4.1, which anchor the spectrin-actin framework to the red cell membrane by attaching to band 3 and glycophorin C.[107] ris mutant embryos manifest profound anemia at day 4, though they can be raised to adulthood. Adult ris maintain a severe anemia, and possess expanded numbers of erythroid precursors in their kidney tissue, as well as cardiomegaly. Peripheral erythrocytes in ris adults are distorted, evincing a spherical or tear-drop morphology with round nuclei. These spherocytic erythrocytes undergo rapid hemolysis associated with compromised marginal band (MB) microtubules, which normally serve to reinforce the red cell membrane in the nonmammallian elliptical erythrocyte. In ris spherocytes, the number of microtubules per MB was found to be reduced by half, suggesting that an intact β-spectrin cytoskeleton is required for aggregation of microtubules into the MB. The blood disorder in ris mutants is analogous to human hereditary spherocytosis (HS), in which patients suffer from hemolytic anemia and their erythroid cells possess an abnormal spherocytic shape.[108]

The mot/cha mutant encodes the zebrafish erythroid specific protein 4.1 (band 4.1/ 4.1R).[105] The ternary complex formed between protein 4.1, spectrin and actin filaments is necessary to maintain stability of the red cell membrane.[109,110] Protein 4.1 deficiency has been shown to result in erythrocytes with an elliptical cell morphology and compromised membrane integrity.[111-113] mot/cha embryos maintain normal erythropoiesis in the first 3 dpf; profound anemia appears at 4 dpf, though a small number (5-10%) of mutant embryos can be raised to adulthood. At 48 hpf, mutant embryo erythrocytes are already morphologically abnormal, possessing spiculated membranes, and a small number of binucleated cells are present in circulation. Examination of the peripheral blood in mot/cha adults found a significant reduction in red cell number; red cells exhibited a differentiation arrest at the basophilic erythroblast stage along with abnormal membrane structure, including surface pitting and membrane projections. Similar to the case in ris fish, the MB is disorganized in mot/cha erythrocytes, reflecting improper assembly or stability of this structure. Adult mot/cha fish suffer severe anemia due to hemolysis accompanied by compensatory kidney hyperplasia. The phenotype of mot/cha corresponds to human patients with hereditary elliptocytosis (HE), a hemolytic anemia characterized by elliptical erythrocytes.[108]

The mot/cha and ris mutants provide genetic models to investigate the genesis of the membrane ultrastructure during erythroid differentiation, as well as provide models of human hemolytic anemias. Human hemolytic anemias are caused by defects in a number of other cytoskeletal proteins, including erythroid α-spectrin, ankyrin, and band 3.[108] Based on ris and mot/cha, the remaining decreasing blood mutants are likely to represent genes encoding these structural components of the erythroid membrane cytoskeleton.

Hypochromic Blood Mutants

The chardonnay (cdy), chianti (cia), clear blood (clb), frascati (fra), sauternes (sau), weissherbst (weh), and zinfandel (zin) mutants evince a phenotype of hypochromic, microcytic anemia, such that primitive erythrocytes are reduced in number and appear both small and abnormally pale.[36,49] Mutants in this class display hypochromic circulation beginning at 33-36 hpf, followed by a marked decrease in the number of circulating erythrocytes. The severity and onset of anemia varies among the mutants, as does the degree of microcytosis (A. Brownlie, A. Donovan, and L.I. Zon, unpublished). Hypochromic, microcytic anemia is commonly indicative of a defect in hemoglobin production during erythroid differentiation, and human disorders of hemoglobin production manifest with these blood characteristics. Hemoglobin production requires both the availability of iron and the de novo synthesis of heme and globin.[114] The cloning of several hypochromic mutants has revealed defects in genes necessary for iron uptake and heme synthesis.

sau mutants have a defect in the erythoid-specific heme enzyme, aminolevulinate synthase 2 (ALAS2), that functions in red cells at the first step in heme biosynthesis.[35] sau embryos display a decrease in cell number at 2 dpf, and primitive erythrocytes differentiate abnormally: they are morphologically immature, maintaining expression of GATA-1 through 72 hpf, as well as misexpressing βe-2 globin at 72 hpf (normally downregulated by 48 hpf). Erythroid heme levels are reduced more than 10-fold in sau embryos, but not absent, consistent with the finding of missense mutations in both sau alleles. sau adults maintain a deficit of heme that is comparatively less severe than that observed in sau embryos, suggesting that reduced alas2 activity is more limiting in the zebrafish embryo. Mutations in ALAS2 cause congenital sideroblastic anemia in humans, and sau represents the first animal model of this disease.

The cdy and weh mutants represent defects in genes responsible for iron uptake in the embryo. cdy is a mutation in the divalent metal transporter (DMT1).[115] The major pathway for iron absorption into a cell utilizes uptake of iron-bound transferrin by transferrin receptor, which subsequently internalizes the iron via clathrin-mediated endocytosis. DMT1 functions to transport iron across the endosome to the cytoplasm in erythroid precursors, and additionally transports iron from the diet across the apical surface of enterocytes in the adult intestine.[116] cdy embryos have a normal number of circulating erythrocytes until 48 dpf, and hemoglobin expression is not detectable. Circulating erythoid cells in cdy embryos are delayed in their differentation based on their immature morphology, though globin gene expressions are normal. cdy mutants are viable, but have an increased number of erythroid precursors present in both their peripheral blood and kidneys, consistent with increased erythropoiesis in response to anemia. The blood phenotype and viability of cdy adults mirrors that of the microcytic anemia mouse (mk) and the Belgrade rat, which harbor mutations in DMT1.[117,118] Based on the adult survival of these DMT1 mutants, it has been proposed that an alternate (albeit minor) protein or pathway functions in vertebrates to intake dietary iron and transport iron into developing erythrocytes, possibly through a mechanism of nontransferrin bound iron uptake.[119,120]

Identification of the weh genetic lesion discovered a novel iron transporter, ferroportin1 (fpn1).[121] weh embryo erythrocytes make virtually no hemoglobin, and similar to the cdy and sau mutants, erythrocytes in circulation are morphologically immature. Intravenous injection of iron-dextran into weh mutant embryos completely rescues hemoglobin production, demonstrating both that weh erythrocytes are capable of hemoglobinization, and that hypochromia in weh mutants is a specific consequence of inadequate iron in the embryonic circulation. fpn1 is expressed between 18-48 hpf in the yolk syncytial layer (the multinuclear yolk cell which surrounds the entire embryo yolk) consistent with a function of transporting iron from maternal yolk stores into embryonic circulation. Examination of fpn1 expression in human and mouse found high levels of fpn1 in the placenta, duodenum, and liver.[121-123] In the placenta, fpn1 was localized to the cell interface between maternal and fetal circulation, analogous to the case in zebrafish embryos. High levels of Ferroportin1 protein at the basolateral surface of duodenal enterocytes suggested that Ferroportin1 functions in the intestine to transport absorbed dietary iron from enterocytes into circulation. Iron export function of Ferroportin1 has since been hypothesized to be involved in the release of iron from hepatocytes and macrophages during iron recycling.[116] Patients with mutations in fpn1 suffer from a rare form of hemachromatosis, an iron overload disorder, in which loss of fpn1 function is believed to cause an imbalance in iron distribution and tissue iron accumulation.[124,125]

The sau, cdy, and weh zebrafish mutants have provided new vertebrate models of hemoglobin production during erythroid differentiation, in each case serving as a disease model for a human hemoglobin production disorder. In particular, cloning of the weh mutant highlights the utility of zebrafish for novel gene discovery as well as the study of iron metabolism. Continued identification of the hypochromic mutants will be useful for establishing genetic pathways for embryonic iron transport, heme biosynthesis, and globin gene expression. At least one mutant, zin, maps to the major globin locus in the zebrafish and is likely to represent a defect in embryonic globin regulation.[35] Additional hypochromic mutants recovered in the Tubingen 2000 large-scale genetic screen promise to serve as additional tools to investigate hemoglobin production during erythroid differentiation (R. Wingert, K. Dooley and L. Zon, unpublished).

Photosensitive Mutants

The dracula (drc), desmodius (dsm), friexenet (frx), and yquem (yqe) mutants produce normal numbers of circulating erythrocytes but the red cells undergo rapid lysis upon exposure to light.[36,49] Along with photosensitivity, the erythrocytes in these mutants also emit autofluorescence prior to lysis.[36,49] This combination of characteristics is consistent with human erythropoeitic porphyria disorders, a group of rare diseases caused by deficiencies in heme biosynthetic enzymes. The heme biosynthetic pathway consists of a cascade of eight enzymes, and defeciencies in the last seven are known to cause human porphyrias.[126] Defects in the last five enzymes of this pathway result in the accumulation of porphyrin intermediates that are toxic to cells: upon exposure to light, the intermediates break down, releasing harmful free radicals which cause red cell lysis.

The drc and yqe genes have been cloned and represent the heme enzymes ferrochelatase (fch) and uroporphyrinogen decarboxylase (UROD), which act at respectively the eighth and fifth steps in heme biosynthesis.[127,128] fch mutations in humans cause erythropoietic protoporphyria (EPP). Along with autoflorescent, light-sensitive red cells, some EPP patients develop liver disease, presumably due to the uptake of toxic substances from lysed erythrocytes. Patients also accumulate crystalline deposits of protoporphyrin in their liver tissue. Similar to humans, dra embryos accumulate high tissue levels of protoporphyrin IX; embryos also develop a red-colored liver with numerous red-brown inclusions scattered throughout the tissue that autofluoresce. drc embryos survive to early larval stages, suggesting either that a maternal source of heme is present in the yolk or that an alternative heme pathway exists. The yqe mutant represents the first genetically accurate model of human hepatoerythopoietic porphyria (HEP). yqe embryos suffer from the ubiquitous UROD deficiency seen in human patients with HEP, rather than a liver-restricted UROD deficiency in chemically induced HEP disease models.[129] yqe embryos accumulate excessive amounts of urophyrinogens I and III and 7-carboxylate porphyrin, consistent with a deficiency of UROD. Enzymatic assays of heterozygous and homozygous yqe embryos showed that UROD activity was reduced 67% and 36%, respectively, as compared to wildtype embryos. yqe mutants die after photo-ablation of their blood, though heterozygous fish appear healthy. Based on the genetic and clinical similarities of the zebrafish heme enzyme mutants to human EPP and HEP, continued study of yqe and drc fish may reveal insight into the pathophysiology of porphyria.

Future Directions

The combined advantages of developmental biology and genetic screens have made the zebrafish an excellent paradigm to study the organogenesis of vertebrate blood. Analysis of the blood mutants created in the large-scale zebrafish screens has revealed insight into a range of the events involved in hematopoiesis. Zebrafish screens have enabled the investigation of complex biological processes without a priori knowledge of the genes involved and have resulted in discovery of many de novo genes. Blood mutants in particular have provided new animal models for several human diseases and are allowing the molecular basis of these diseases to be uncovered. The ongoing zebrafish genome initiative will provide a sequenced genome in the near future, which will greatly facilitate genetic cloning.

Continued study of the existing blood mutants, along with the implementation of future screens, promises a wealth of genetic insight into hematopoiesis. There remain a number of early genes for which a zebrafish mutant has not been found, including scl, lmo2, and GATA-2. In addition, mutants with specific myeloid, lymphoid, or thrombocyte deficiencies were not identified. More recently, a second large scale screen was conducted in Tubingen, Germany and has yielded a number of new blood mutants currently under study in our lab. Future screens which target a single pathway represent the next step in using the zebrafish to study hematopoiesis, and a wide range of inventive screening strategies have been developed by zebrafish researchers.[130] Among these, screen schemas that evaluate gene expression by in situ hybridization or by utilizing transgenic zebrafish lines will allow specific aspects of hematopoiesis to be examined.[131] A recent screen in our laboratory, for example, has isolated a number of mutants with lymphoid lineage defects after screening for zebrafish embryos with absent rag-1 expression by in situ hybridization.[38] Insertional mutagenesis has proven an effective strategy to create mutants with developmental defects and rapidly determine the disrupted genes.[132] Suppressor/enhancer screens in zebrafish will facilitate further identification of genes required for hematopoiesis.

Complementary to genetic screen strategies, in situ hybridization screening of cDNAs collected from various zebrafish libraries has been used to identify and characterize hematopoietic genes expressed during embryogenesis.[19] Functional analysis of genes is now possible in zebrafish using antisense morpholino oligonucleotides (MOs), which knockdown gene function by either preventing translation or altering transcriptional processing.[133,134] MOs have been used against a diverse range of genes, including those involved in hematopoiesis, and successfully phenocopy zebrafish genetic mutants.[133-136] A reverse genetics approach in zebrafish, involving random chemical mutagenesis followed by targeted screening for induced mutations in the rag1 gene, has been implemented successfully to obtain rag1 mutants.[137] This approach enables the identification of mutant alleles for any gene of interest.

The diversity of tools available to zebrafish geneticists provides unique opportunities for the use of this exciting model system in the study of hematopoiesis. Forward genetics will continue to find new genes that regulate stem cell homeostasis and blood differentiation. Reverse genetics will create mutants that could be used in suppressor/enhancer screens to extensively dissect the genetic pathways required for blood formation. Functional genomic analyses of gene expression during embryogenesis should provide a useful resource of candidates for the genetic mutants created in future screens. It is possible that marrow transplantation studies will be investigated at a genetic level in the zebrafish. The combination of developmental and cell biology, coupled with genetics, should make the zebrafish an invaluable vertebrate system for future studies of human blood diseases and leukemias.

Acknowledgements

We thank Alan Davidson for providing (Fig. 1), Jenna Galloway, Noelle Paffett-Lugassy, and David Traver for providing (Fig. 2), and Dorothy Giarla for providing (Fig. 4) We thank Jenna Galloway for critical review of this manuscript, and Alan Davidson, Kim Dooley, Noelle Paffett-Lugassy and Barry Paw for helpful discussions. R.A.W. is supported by Hematology Training Grant, T32 HL07623. L.I.Z. is an Associate Investigator of the Howard Hughes Medical Institute.

References

1. Amatruda JF, Zon LI. Dissecting hematopoiesis and disease using the zebrafish. Dev Biol 1999; 216:1-15.
2. Driever W, Fishman MC. The zebrafish: heritable disorders in transparent embryos. J Clin Invest 1996; 97:1788-1794.
3. Dooley K, Zon LI. Zebrafish: a model system for the study of human disease. Curr Opin Gen Dev 2000; 10:252-256.
4. Ward AC, Lieschke GJ. The zebrafish as a model system for human disease. Front Biosci 2002; 7:827-833.
5. Brownlie A, Zon LI. The zebrafish as a model system for the study of hematopoiesis. BioScience 1999; 49:382-392.
6. Paw BH, Zon LI. 2000. Zebrafish: a genetic approach in studying hematopoiesis. Curr Opin Hematol 2000; 7:79-84.
7. Amemiya CT, Zon LI. Generation of a zebrafish P1 artificial chromosome library. Genomics 1999; 58:211-213.
8. Geisler R, Rauch G, Baier H et al. A radiation hybrid map of the zebrafish genome. Nat Genet 1999; 23:86-89.
9. Hukriede NA, Joly L, Tsang M et al. Radiation hybrid mapping of the zebrafish genome. Proc Natl Acad Sci USA 1999; 96:9745-9750.
10. Knapik EW, Goodman A, Atkinson OS et al. A reference cross DNA panel for zebrafish (Danio rerio) anchored with simple sequence length polymorphism. Development 1996; 123:451-460.
11. Knapik EW, Goodman A, Ekker M et al. A microsatellite genetic linkage map for zebrafish (Danio rerio). Nat Genet 1998; 18:338-343.
12. Postlewait JH, Yan YL, Gates M et al. Vertebrate genome evolution and the zebrafish gene map. Nat Genet 1998; 18:345-349.

13. Shimoda N, Knapik EW, Ziniti J et al. Zebrafish genetic map with 2000 microsatellite markers. Genomics 1999; 58:219-232.
14. Zhong TP, Kaphigst K, Akella U et al. Zebrafish genomic library in yeast artifical chromosomes. Genomics 1998; 48:136-138.
15. Al-Adhami MA, Kunz YW. Ontogenesis of hematopoietic sites in the Brachydanio rerio. Dev Growth Differ 1977; 19:171-179.
16. Zon LI. Developmental biology of hematopoiesis. Blood 1995; 86:2876-2891.
17. Detrich HW, Kieran MW, Chan FY et al. Intraembryonic hematopoietic cell migration during vertebrate development. PNAS 1995; 92:10713-10717.
18. Willett CE, Cortes A, Zuasti A et al. Early hematopoiesis and developing lymphoid organs in the zebrafish. Dev Dyn 1999; 214:323-336.
19. Herbomel P, Thisse B, Thisse C. Ontogeny and behavior of early macrophages in the zebrafish embryo. Development 1999; 126:3735-3745.
20. Bennett C, Kanki JP, Rhodes J et al. Myelopoiesis in the zebrafish, Danio rerio. Blood 2001; 98:643-651.
21. Lieschke G, Oates AC, Crowhurst MO et al. Morphologic and functional characterization of granulocytes and macrophages in embryonic and adult zebrafish. Blood 2001; 98:3087-3096.
22. Brownlie A, Hersey C, Oates AC et al. Characterization of embryonic globin genes of the zebrafish. Dev Biol 2003; 255:48-61.
23. Catton, WT. Blood cell formation in certain teleost fishes. Blood 1951; 6:39-60.
24. Rowley AF, Hunt TC, Page M et al. In: Vertebrate Blood Cells. Cambridge University Press, Cambridge, 1988; 19-128.
25. Chan FY, Robinson J, Brownlie A et al. Characterization of adult α- and β-globin genes in the zebrafish. Blood 1997; 89:688-700.
26. Gering M, Rodaway ARF, Gottgens B et al. The SCL gene specifies hemangioblast development from early mesoderm. EMBO 1998; 17:4029-4045.
27. Kalev-Zylinska ML, Horsfield JA, Flores MV et al. Runx1 is required for zebrafish blood and vessel development and expression of a human RUNX1-CBF2T1 transgene advances a model for studies of leukemogenesis. Development 2002; 129:2015-2030.
28. Liao EC, Paw BH, Oates AC et al. SCL/Tal-1 transcription factor acts downstream of cloche to specify hematopoietic and vascular progenitors in the zebrafish. Genes Dev 1998; 12:621-626.
29. Liao EC, Paw BH, Peters LL et al. Hereditary spherocytosis in zebrafish riesling illustrates evolution of erythroid B-spectrin structure, and function in red cell morphogenesis and membrane stability. Development 2000; 127:5123-5132.
30. Lieschke G, Oates AC, Paw BH et al. Zebrafish SPI-1 (PU.1) marks a site of myeloid development independent of primitive erythropoiesis: implications for axial patterning. Dev Biol 2002; 246:274-295.
31. Thompson MA, Ransom DG, Pratt SJ et al. The cloche and spadetail genes differentially affect hematopoiesis and vasculogenesis. Dev Biol 1998; 197:248-269.
32. Willett CE, Cherry JJ, Steiner LA. 1997a. Characterization and expression of the recombination activating genes (rag1 and rag2) of zebrafish. Immunogenetics 1997; 45:394-404.
33. Willett CE, Kawasaki H, Amemiya CT et al. Ikaros expression as a marker for lymphoid progenitors during zebrafish development. Dev Dyn 2001; 222:694-698.
34. Liao W, Ho CY, Yan YL et al. Hhex and scl function in parallel to regulate early endothelial and blood differentiation in zebrafish. Development 2000; 127:4303-4313.
35. Brownlie A, Donovan A, Pratt SJ et al. Positional cloning of the zebrafish sauternes gene: a model for congenital sideroblastic anaemia. Nat Genet 1998; 20:244-250.
36. Weinstein BM, Schier AF, Abdelilah S et al. 1996. Hematopoietic mutations in the zebrafish. Development 1996; 123:303-309.
37. Willett CE, Zapata AG, Hopkins N et al. Expression of zebrafish rag genes during early development identifies the thymus. Dev Biol 1997; 182:331-341.
38. Trede NS, Zapata A, Zon LI. Fishing for lymphoid genes. Trends Immunol 2001; 22:302-307.
39. Parichy DM, Ransom DG, Paw B et al. An orthologue of the kit-related gene fms is required for development of neural crest-derived xanthophores and a subpopulation of adult melanocytes in the zebrafish, Danio rerio. Development 2000; 127:3031-3044.

40. Shepard JL, Zon LI. Developmental derivation of embryonic and adult macrophages. Curr Opin Hematol 2000; 7:3-8.
41. Allende M, Amsterdam A, Becker T et al. Insertional mutagenesis in zebrafish identifies two novel genes, pescadillo and dead eye, essential for embryonic development. Genes Dev 1996; 10:3141-3155.
42. Amsterdam A, Burgess S, Golling G et al. A large-scale insertional mutagenesis screen in zebrafish. Genes Dev 1999; 3:2713-2724.
43. Bahary N, Zon LI. Use of the zebrafish (Danio rerio) to define hematopoiesis. Stem Cells 1998; 16:89-98.
44. Gaiano N, Amsterdam A, Kawakami K et al. Insertional mutagenesis and rapid cloning of essential genes in zebrafish. Nature 1996; 383:829-832.
45. Mullins M, Hammerschmidt M, Haffter P et al. Large-scale mutagenesis in the zebrafish: in search of genes controlling development in a vertebrate. Curr Biol 1994; 4:189-202.
46. Solnica-Krezel L, Schier A, Driever W. Efficient recovery of ENU-induced mutations from the zebrafish germline. Genetics 1994; 136:1401-1420.
47. Driever W, Solnica-Krezel L, Schier AF et al. A genetic screen for mutations affecting embryogenesis in zebrafish. Development 1996; 123:37-46.
48. Haffter P, Granato M, Brand M et al. The identification of genes with unique and essential functions in the development of the zebrafish, Danio rerio. Development 1996; 123:1-36.
49. Ransom DG, Haffter P, Odenthal J et al. Characterization of zebrafish mutants with defects in embryonic hematopoiesis. Development 1996; 123:311-319.
50. Kimmel CB, Kane DA, Walker C et al. A mutation that changes cell movement and cell fate in the zebrafish embryo. Nature 1989; 337:358-362.
51. Stainier DYR, Weinstein BM, Deitrich HW et al. cloche, an early acting zebrafish gene, is required by both the endothelial and hematopoietic lineages. Development 1995; 121:3141-3150.
52. Liao EC, Trede NS, Ransom D et al. Noncell autonomous requirement for the bloodless gene in primitive hematopoiesis of zebrafish. Development 2002; 129:649-659.
53. Davidson AJ, Zon LI. Turning mesoderm into blood: the formation of hematopoietic stem cells during embryogenesis. Curr Top Dev Biol 2000; 50:45-60.
54. Nieto MA. Reorganizing the organizer 75 years on. Cell 1999; 98:417-25.
55. Dale L, Wardle FC. A gradient of BMP activity specifies dorsal-ventral fates in early Xenopus embryos. Sem Cell Dev Bio 1999; 10:319-326.
56. Baker JC, Harland RM. From receptor to nucleus: the Smad pathway. Curr Opin Gen Dev 1997; 7:467-473.
57. Harland R, Gerhart J. Formation and function of Spemannís organizer. Ann Rev Cell Dev Biol 1997; 13:611-667.
58. Mullins M, Hammerschmidt M, Kane DA et al. Genes establishing dorsoventral pattern formation in the zebrafish embryo: the ventral specifying genes. Development 1996; 123:81-93.
59. Lee KH, Marden JJ, Thompson MS et al. Cloning and genetic mapping of zebrafish BMP-2. Dev Genetics 1998; 23:97-103.
60. Dick A, Hild M, Bauer H et al. Essential role of bmp7 (snailhouse) and its prodomain in dorsoventral patterning of the zebrafish embryo. Development 2000; 127:343-354.
61. Schmid B, Furthauer M, Connors SA et al. Equivalent roles for bmp7/snailhouse and bmp2b/swirl in dorsoventral pattern formation. Development 2000; 127:957-967.
62. Hild M, Dick A, Rauch GJ et al. The smad5 mutation somitabun blocks Bmp2b signaling during early dorsoventral patterning of the zebrafish embryo. Development 1999; 126:2149-2159.
63. Bauer H, Lele Z, Rauch GJ et al. The type 1 serine/threonine kinase receptor Alk8/Lost-a-fin is required for Bmp2b/7 signal transduction during dorsoventral patterning of the zebrafish. Development 2001; 128:849-858.
64. Mintzer KA, Lee MA, Runke G et al. lost-a-fin encodes a type 1 BMP receptor, Alk8, acting maternally and zygotically in dorsoventral pattern formation. Development 2001; 128:859-869.
65. Hammerschmidt M, Pelegri F, Mullins MC et al. dino and mercedes, two genes regulating dorsal development in the zebrafish embryo. Development 1996; 123:95-102.
66. Schulte-Merker S, Lee KJ, McMahon AP et al. The zebrafish organizer requires chordino. Nature 1997; 387:862-863.

67. Connors SA, Trout J, Ekker M et al. The role of tolloid/mini fin in dorsoventral pattern formation of the zebrafish embryo. Development 1999; 126:3119-3130.
68. Blader P, Rastegar S, Fischer N et al. Cleavage of the BMP-4 antagonist chordin in zebrafish tolloid. Science 1997; 278:1937-1940.
69. Miller-Bertolio V, Carmany-Rampey A, Furthauer M et al. Maternal and zygotic activity of the zebrafish ogon locus antagonizes BMP signaling. Dev Biol 1999; 214:72-86.
70. Wagner DS, Mullins MC. Modulation of BMP activity in dorsal-ventral pattern formation by the chordin and ogon antagonists. Dev Biol 2002; 245:109-123.
71. Zhang J, King ML, Houston D et al. The role of maternal VegT in establishing the primary germ layers in Xenopus embryos. Cell 1998; 94:515-524.
72. Ho RK, Kane DA. Cell-autonomous action of zebrafish spt-1 mutation in specific mesodermal precursors. Nature 1990; 388:728-730.
73. Griffin KJP, Amacher SL, Kimmel CB et al. Molecular identification of spadetail: regulation of zebrafish trunk and tail mesoderm formation by T-box genes. Development 1998; 125:3379-3388.
74. Murray PDF. The development in vitro of the blood of the early chick embryo. Proc Roy Soc London 1932; 11:497-521.
75. Sabin FR. Studies on the origin of blood vessels and of red blood corpuscles as seen in the living blastoderm of chicks during the second day of incubation. Contrib Embryol 1920; 9:213-262.
76. Pardanaud L, Altmann C, Kitos P et al. Vasculogenesis in the early quail blastodisc as studied with a monoclonal antibody recognizing endothelial cells. Development 1987; 100:339-349.
77. Pardanaud L, Yassine F, Dieterlen-Lievre D. Relationship between vasculogenesis, angiogenesis, and hemapoiesis during avian ontogeny. Development 1989; 105:473-485.
78. Pardanaud L, Luton D, Prigent M et al. Two distinct endothelial lineages in ontogeny, one of them related to hemopoiesis. Development 1996; 122:1363-1371.
79. Lacaud G, Robertson S, Palis J et al. Regulation of hemangioblast development. Ann NY Acad Sci 2001; 938:96-107.
80. Bernex F, De Sepulveda P, Kress C et al. Spatial and temporal patterns of c-kit expressing cells in the WlacZ/+ and Wlacz/Wlacz mouse embryos. Development 1996; 122:3023-3033.
81. Drake CJ, Brandt SJ, Trusk TC et al. Tal1/SCL is expressed in endothelial progenitor cells/ angioblasts and defines a dorsal-to-ventral gradient of vasculogenesis. Dev Biol 1997; 192:17-30.
82. Iwama A, Hamaguchi I, Hashiyma M et al. Molecular cloning and characterization of mouse TIE and TEK receptor tyrosine kinase genes and their expression in hematopoietic stem cells. Biochem Biophys Res Commun 1993; 195:301-309.
83. Kabrun N, Buhring HJ, Choi K et al. Flk-1 expression defines a population of early embryonic hematopoietic precursors. Development 1997; 124:2039-2048.
84. Kallianpur AR, Jordan JE, Brandt SJ. The SCL/TAL-1 gene is expessed in progenitors of both the hematopoietic and vascular systems during embryogenesis. Blood 1994; 83:1200-1208.
85. Shalaby F, Ho J, Stanford KD et al. A requirement for flk1 in primitive and definitive hematopoiesis and vasculogenesis. Cell 1997; 89:981-990.
86. Wood HB, May G, Healy L et al. CD34 expression patterns during early mouse development are related to modes of blood vessel formation and reveal additional sites of hematopoiesis. Blood 1997; 90:2300-2311.
87. Yamaguchi TP, Dumont DJ, Conlon RA et al. flk-1, a flt-related receptor tyrosine kinase is an early marker for endothelial cell precursors. Development 1993; 118:489-498.
88. Young P, Baumhueter S, Lasky L. The sialomucin CD34 is expressed on hematopoietic cells and blood vessels during murine development. Blood 1995; 85:96-105.
89. Shalaby F, Rossant J, Yamaguchi TP et al. Failure of blood-island formation and vasculogenesis in flk-1 deficient mice. Nature 1995; 376:62-66.
90. Okuda T, vanDeursen J, Hiebert SW et al. AML1, the target of multiple chromosomal translocations in human leukemia, is essential for normal fetal liver hematopoiesis. Cell 1996; 84:321-300.
91. Wang Q, Stacy T, Miller JD et al. The CBFb subunit is essential for CBPa2 (AML1) function in vivo. Cell 1996; 87:697-708.
92. Kennedy M, Firpo M, Choi K et al. A common precursor for primitive erythropoiesis and definitive hematopoiesis. Nature 1997; 386:488-493.

93. Choi K, Kennedy M, Kazarov A et al. A common precursor for hematopoietic and endothelial cells. Development 1998; 125:725-732.

94. Robb L, Elefanty AG. The hemangioblast—an elusive cell captured in culture. BioEssays 1998; 611-614.

95. Liao W, Bisgrove BW, Sawyer H et al. The zebrafish gene cloche acts upstream of a flk-1 homologue to regulate endothelial cell differentiaion. Development 1997; 124:381-389.

96. Porcher C, Liao EC, Fujiwara Y et al. Specification of hematopoietic and vascular development by the bHLH transcription factor SCL without direct DNA binding. Development 1999; 126:4603-4615.

97. Parker L, Stainier DYR. Cell-autonomous and nonautonomous requirements for the zebrafish gene cloche in hematopoiesis. Development 1999; 126:2643-2651.

98. Lyons SE, Lawson ND, Lei L et al. A nonsense mutation in zebrafish gata1 causes the bloodless phenotype in vlad tepes. Pro Natl Acad Sci USA 2002; 99:5454-5459.

99. Fujiwara Y, Browne CP, Cunniff K et al. Arrested development of embryonic red cell precursors in mouse embryos lacking transcription factor GATA-1. PNAS 1996; 93:12355-12358.

100. Pevney L, Simon MC, Robertson E et al. Erythroid differentiation in chimaeric mice blocked by a targeted mutation in the gene for transcription factor GATA-1. Nature 1991; 349:257-260.

101. Pevney L, Lin CS, DíAgati V et al. Development of hematopoietic cells lacking transcription factor GATA-1. Development 1995; 121:163-172.

102. Weiss MJ, Keller G, Orkin SH. Novel insights into erythroid development revealed through in vitro differentiation of GATA-1 embryonic stem cells. Genes Dev 1994; 8:1184-1197.

103. Kelsh RN, Brand M, Jiang YJ et al. Zebrafish pigmentation mutations and the processes of neural crest development. Development 1996; 123:369-389.

104. Parichy DM, Rawls JF, Pratt SJ et al. Zebrafish sparse corresponds to an orthologue of c-kit and is required for the morphogenesis of a subpopulation of melanocytes, but is not essential for hematopoiesis or primordial germ cell development. Development 1999; 126:3425-3436.

105. Shafizadeh E, Paw BH, Foott H et al. Characterization of zebrafish merlot/chablis as a nonmammalian vertebrate model for severe congenital anemia due to protein 4.1 deficiency. Development 2002; 129:4359-4370.

106. Dubreuil RR, Grushko T. Genetic studies of spectrin: new life for a ghost protein. BioEssays 1998; 20:875-878.

107. DeMatteis MA, Morrow JS. Spectrin tethers and mesh in the biosynthetic pathway. J Cell Sci 2000; 113:2331-2343.

108. Lux SE, Palek J. Disorders of the red cell membrane. in Blood: Principles and practice of hematology JB Loppincott Co, Philadelphia, 1995; 1701-1818.

109. Conboy JG. Structure, function, and molecular genetics of erythroid membrane skeletal protein 4.1 in normal and abnormal red blood cells. Semin Hematol 1993; 30:58-73.

110. Lorenzo F, Dalla VN, Morle L et al. Protein 4.1 deficiency associated with an altered binding to the spectrin-actin complex of the red cell membrane skeleton. J Clin Invest 1994; 94:1651-1656.

111. Marchesi SL, Conboy J, Agre P et al. Molecular analysis of insertion/deletion mutations in elliptocytosis. I. Biochemical identification of rearrangements in the spectrin/actin binding domain and functional characterizations. J Clin Invest 1990; 86:516-523.

112. Shi ZT, Afzal V, Coller B et al. Protein 4.1R-deficient mice are viable but have erythroid membrane skeleton abnormalities. J Clin Invest 11999; 03:331-340.

113. Yawata A, Kanzaki A, Gilsanz F et al. A markedly disrupted skeletal network with abnormally distributed intramembrane particles in complete protein 4.1-deficient red blood cells (allele 4.1 Madrid): implication regarding a critical role of protein 4.1 in maintenance of the integrity of the red cell membrane. Blood 1997; 90:2471-2481.

114. Lux SE. Introduction to anemias. In Blood: Principles and practice of hematology. Philadelphia: JB Loppincott Co, 1995:1383-1398.

115. Donovan A, Brownlie A, Dorschner MO et al. The zebrafish mutant gene chardonnay encodes Divalent Metal Transporter1 (DMT1). Blood 2002; 100:4655-4659.

116. Andrews NC. Iron homeostasis: insights from genetics and animal models. Nat Rev Genetics 2000; 1:208-217.

117. Fleming MD, Trenor CC, Su MA et al. Microcytic anaemia mice have a mutation in Nramp2, a candidate iron transporter gene. Nat Genet 1997; 16:383-386.
118. Fleming MD, Romano MA, Su MA et al. Nramp2 is mutated in the anemic Belgrade (b) rat: evidence of a role for Nramp2 in endosomal iron transport. Proc Natl Acad Sci USA 1998; 95:1148-1153.
119. Gelvan D, Fibach E, Meyron-Holtz EG et al. Ferritin uptake by human erythroid precursors is a regulated iron uptake pathway. Blood 1996; 88:3200-3207.
120. Meyron-Holtz EG, Vaisman B, Cabantchik ZI et al. Regulation of intracellular iron metabolism in human erythroid precursors by internalized extracellular ferritin. Blood 1999; 94:3205-3211.
121. Donovan A, Brownlie A, Zhou Y et al. Positional cloning of zebrafish ferroportin1 identifies a conserved vertebrate iron exporter. Nature 2000; 403:776-781.
122. Abboud S, Haile DJ. A novel mammalian iron-regulated protein involved in intracellular iron metabolism. J Biol Chem 2000; 275:19906-19912.
123. McKie AT, Marciani P, Rolfs A et al. A novel duodenal iron-regulated transporter, IREG1, implicated in the basolateral transfer of iron to the circulation. Mol Cell 2000; 5:299-309.
124. Montosi G, Donovan A, Totaro A et al. Autosomal-dominant hemochromatosis is associated with a mutation in the ferroportin (SLC11A3) gene. J Clin Invest 2001; 108:619-623.
125. Njajou OT, Vaessen N, Joosse M et al. A mutation in SLC11A3 is associated with autosomal dominant hemochromatosis. Nat Genet 2001; 28:213-214.
126. Kappas A, Sassa S, Galbraith RA et al. In: The Metabolic Basis of Inherited Diseases. New York: McGraw-Hill, 1995:2103-2159.
127. Childs S, Weinstein BM, Mohideen MPK et al. Zebrafish dracula encodes ferrochelatase and its mutation provides a model for erythorpoietic protoporphyria. Curr Biol 2000; 10:1001-1004.
128. Wang H, Long Q, Marty SD et al. A zebrafish model for hepatoerythropoietic porphyria. Nat Genetics 1998; 20:239-243.
129. Elder GH, Roberts AG, de Salamanca RE et al. Genetics and pathogenesis of human uroporphyrinogen decarboxylase defects. Clin Biochem 1989; 22:163-168.
130. Patton EE, Zon LI. The art and design of genetic screens: zebrafish. Nat Gen Rev 2001; 2:956-966.
131. Long Q, Meng A, Wang H et al. GATA-1 expression pattern can be recapitulated in living transgenic zebrafish using GFP reporter gene. Development 1997; 124:4106-4111.
132. Golling G, Amsterdam A, Zhaoxia S et al. Insertional mutagenesis in zebrafish rapidly identifies genes essential for early vertebrate development. Nat Genet 2002; 31:135-140.
133. Nasevicius A, Ekker SC. Effective targeted gene knockdown in zebrafish. Nat Genet 2000; 26:216-220.
134. Ekker SC, Larson JD. Morphant technology in model developmental systems. Genesis 2001; 30:89-93.
135. Lele Z, Bakkers J, Hammerschmidt M. Morpholino phenocopies of the swirl, snailhouse, somitabun, minifin, silberblick, and pipetail mutations. Genesis 2001; 30:190-194.
136. Imai Y, Talbot WS. Morpholino phenocopies of the bmp2b/swirl and bmp7/snailhouse mutations. Genesis 2001; 30:160-163.
137. Wienholds E, Shulte-Merker S, Walderich B et al. Target-selected inactivation of the zebrafish rag1 gene. Science 2002; 297:99-102.

CHAPTER 3

Extra- and Intraembryonic HSC Commitment in the Avian Model

Thierry Jaffredo, Karine Bollerot, Krisztina Minko, Rodolphe Gautier, Stéphane Romero and Cécile Drevon

Introduction

Hematopoietic stem cells (HSC) are at the basis of the hematopoietic system construction. In adult higher Vertebrates, HSC, defined by their multipotentiality and self-renewal capacity, settle in the bone marrow where they can differentiate into progenitors with more restricted lineage potential and generate all blood lineages via a cascade of commitment events. However HSC are generated during the earliest phases of embryonic development into specific sites. Genetic technologies in the mouse have revealed a number of mutations that affect the production of blood cells, some of which early during development. The tiny mouse embryo embedded in the uterus is not however the most appropriate model to study the earliest events of the development for the analysis of cell commitment, cell migration and cell interaction. Work in the avian embryo has led to several breakthroughs in analysing the ontogeny of the hematopoietic system. Here we will review the main steps that have paved a 30 year analysis of the construction of the hematopoietic system.

The HSC That Colonise the Hematopoietic Organs Have an Extrinsic Origin

One fundamental aspect, first established in the avian model and confirmed thereafter in all vertebrate species, is that hematopoietic organ rudiments are colonised by extrinsic HSC. This notion was first envisioned by Moore and Owen on the basis of observations on parabiosed embryos or chorioallantoic grafts of organ rudiments.[44,45] The marker used was the pair of heterochromosomes, typical of the female sex in chicks, which became identifiable in colchicine-treated cells arrested in metaphase. These two authors led to the conclusion that the hematopoietic organ rudiments, including the mammalian fetal liver, bone marrow and spleen were colonized by cells coming from elsewhere. This has made the founder experiments of the hematogenous theory.[46] These conclusions were confirmed and extended, with the quail chick/ marker leading to the fine-tuning dissection of thymic colonisation.[30,38,39]

The Discovery of an Intraembryonic Hematopoietic Site

In the same set of experiments, Moore and Owen also postulated a central role for the yolk sac (YS) in the production of HSC. According to the authors, this appendage produced a stock of HSC, a part of which would differentiate in situ while the other part would colonize the hematopoietic organ rudiments.[46] The quail/chick system combined with the accessibility and topology of the avian embryo, which lay flat upon the spherical yolk sac, made possible the identification of the relative contributions of the extraembryonic and intraembryonic compartments in the production of HSC. This elegant experiment, designed by F. Dieterlen and coworkers, consisted of associating a quail embryo onto a chick yolk sac and analysing the constitution of the hematopoietic organs for quail or chick cells.[20] Hematopoietic organs were seeded exclusively by quail cells. Contrary to the centripetal flow of hematopoietic cells (HC) postulated by Moore and Owen, cells from the embryo even seeded the YS by Embryonic day 5 (E5).[4,40] At E13, 70% of circulating blood cells were of quail origin.[4] Similar experiments were repeated with congenic strains of chicks differing either by their sex chromosomes,[35] their immunoglobulin haplotypes[35,41] or major histocompatibility antigens[37] leading to the same conclusions. YS cells decreased rapidly from E5 to be absent at hatching. Thus, yolk sac hematopoietic cells are incapable of long-term renewal whereas the source of "definitive" HSC colonising the hematopoietic organs comes from the embryo proper. These experiments also revealed that the YS contributes to two distinct generations of red cells: a primitive one that derives entirely from the YS and a second one (fetal) that derives both from the embryo and the yolk sac, the latter being able to generate erythrocytes with a fetal globin make up.[4,5] The restricted differentiation capacities of YS HSC probably reflect the influence of the YS microenvironment that may trigger HSC differentiation into erythroid and myeloid lineages. Of note is the fact that if an extraembryonic area of an E2 quail embryo is associated in vitro with an E6 chick thymus, quail HSC colonize the thymus rudiment and undergo lymphoid differentiation[29] meaning that this potentiality is present but masked by the microenvironment. Finally, using inversed complementary chimeras (chick embryo grafted onto a quail YS) combined with the use of the cell type- and species-specific monoclonal antibodies QH1/MB1[50,53] recognizing quail hemopoietic and endothelial cells, it was shown that YS produced a small population of HC with characteristics of primitive macrophages, found in zones of apoptosis, which may represent the first microglial cells.[17,18] Such a YS contribution was also recently demonstrated for the mouse embryo.[2]

Where do intraembryonic HSC emerge? The aortic region, also called the p-Sp/AGM (para-aortic Splanchnopleura/Aorta, Gonad, Mesonephros that consists of the aorta and the surrounding tissues at two successive stages of development) in the case of mouse and human embryos, was shown to harbour HSC. In the avian embryo, the aorta exhibits two different aspects of hematopoiesis; one occurring between E3 and E4, called the intraaortic clusters, which consists in small groups of cells with hematopoietic characters protruding into the aortic lumen, a second, called the paraaortic foci, developing between E7 and E9 outside the aorta and consisting in large groups of cells distributed in a loose mesenchyme ventral to the vessel.[21] Whether or not the second aspect develops in mouse and human embryos is not yet known. Theses two sites were demonstrated to harbour transplantable HSC. Isolated E4 chick or quail aortae grafted onto a reverse host produced HC capable of seeding the hematopoietic organs and acquired surface antigens characteristics of T or B lymphocytes and granulocytes.[13] When dissociated into individual cells and seeded into a semi-solid medium supplemented with appropriate cytokines, chick aortic cells displayed erythroid, monocytic, myeloid or multipotential colonies.[14-16]

Gastrulation and Patterning of the Mesoderm

The blood-forming system i.e., endothelial and hematopoietic cells, differentiate from the mesoderm during the gastrulation process i.e., the ingression of cells from the superficial epiblast. Extraembryonic mesoderm is deposited first and forms the area opaca (AO), the

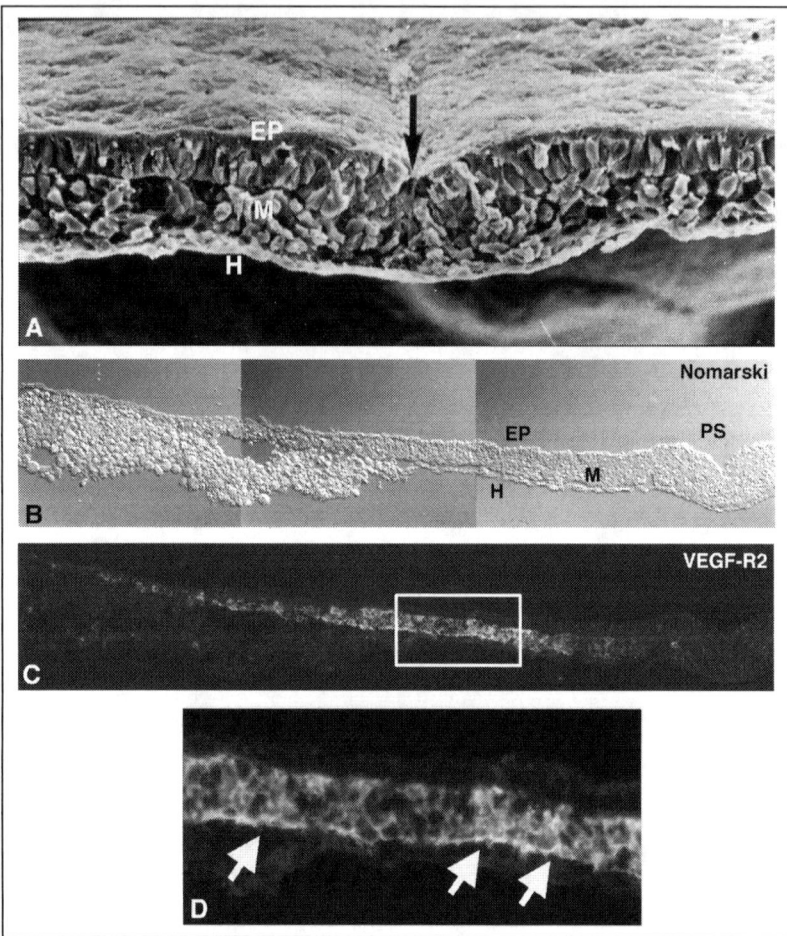

Figure 1. Gastrulation. Definitive streak stage (18h of incubation). A) Scanning electron micrograph. Cross section through the primitive streak. The arrow indicates the primitive streak. Cells ingressing through the streak are forming the mesoderm, the intermediate cell layer. B) Cross section through the posterior part of the primitive streak. Nomarski's interference contrast. C) Same section stained for VEGF-R2 protein expression. A large number of mesodermal cells express the receptor. D) High magnification of the frame in (C). Arrows point out a layer of cells that displays higher levels of the receptor. EP: epiblast; M: mesoderm; H: hypoblast; PS: primitive streak.

region lateral to the embryo proper. The AO is so-called because cells constituting the endodermal layer, the endoblast, are yolky thus looking opaque. Mesoderm giving rise to blood islands (BI) was shown to be deposited at early stages from cells located in the posterior part of the primitive streak.[54] Cells first deposited undergo extensive lateralwards and cephalwards migrations and reach the edges of the blastodisc, while cells passing through the streak immediately after occupy more medial and more caudal positions. Thereafter, cells gradually fill more central and more posterior positions as the primitive streak regresses from cephalic to caudal levels. Thus, early deposited cells have a longer life history than their immediate neighbours and undergo differentiation earlier. BI wherefrom the first hematopoietic cells are going to

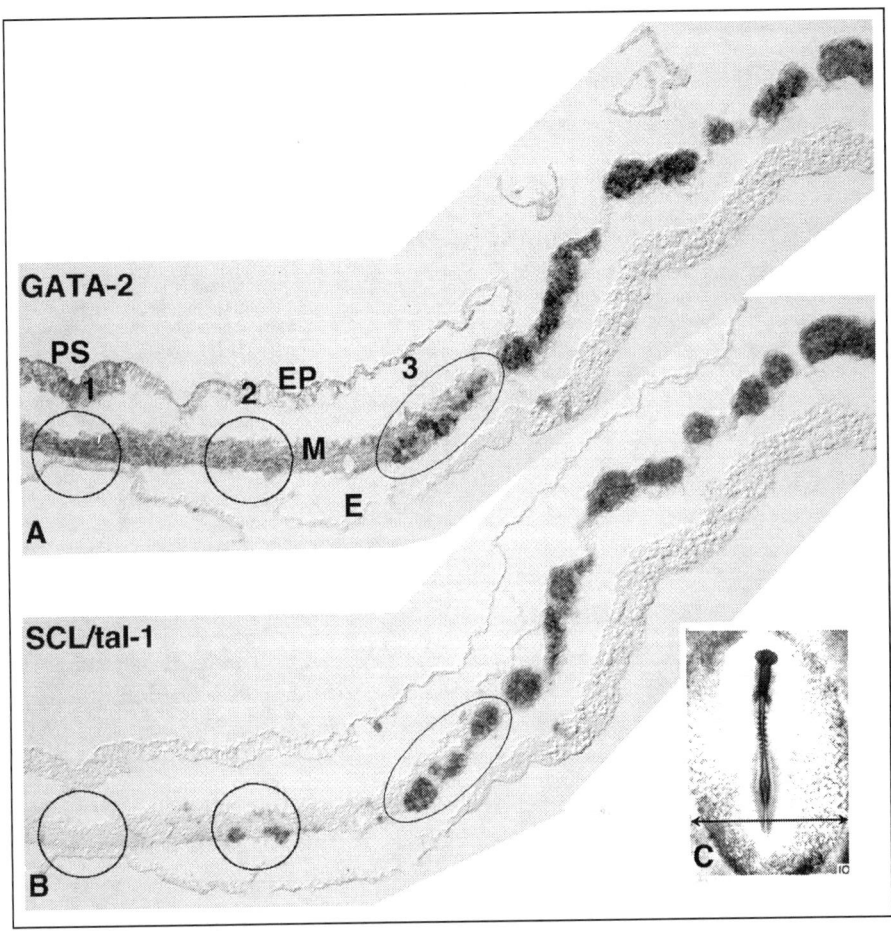

Figure 2. Formation of the blood islands. A), B) Embryo of ten pairs of somites. Adjacent cross sections through the primitive streak. In situ hybridisation with GATA-2 (A) and SCL/Tal-1 (B) probes. The circles, numbered from 1 to 3, point out specific steps in the formation of blood islands from the earliest in the left side to the latest in the right side. A) Steps 1 and 2. The whole ingressing mesoderm and the epiblast appear weakly positive for GATA-2. GATA-2 is upregulated from step 3 when blood island-forming cells become packed together forming a conspicuous structure. In the next steps, a high GATA-2 expression was maintained. B) Step 1. The whole mesoderm is negative for SCL/Tal-1. Step 2. The first SCL/Tal-1+ cells are small clusters of hemangioblastic cells close to the endoderm. Note that these clusters did not upregulate GATA-2. Step 3. Blood islands, completely positive for SCL/Tal-1, becomes conspicuous. Note that the island is not completely GATA-2 + at that stage. C) 10 somite pair embryo. The arrow indicates the level of the section.

emerge, differentiate from the AO mesoderm. BI are typically constituted of an outer layer of endothelial cells and a core of hematopoietic cells. This region will progressively spread onto the entire YS. Mesoderm giving rise to the embryonic area, designated as the Area Pellucida AP; so-called because this area lays above the segmentation cavity thus appears clear, is generated by cells occupying a more anterior position along the primitive streak.[54] At the time BI differentiate, the AP wherefrom the embryo proper develops does not participate in this early blood-forming phase, however it contains structures very similar to YS BI (that will be referred

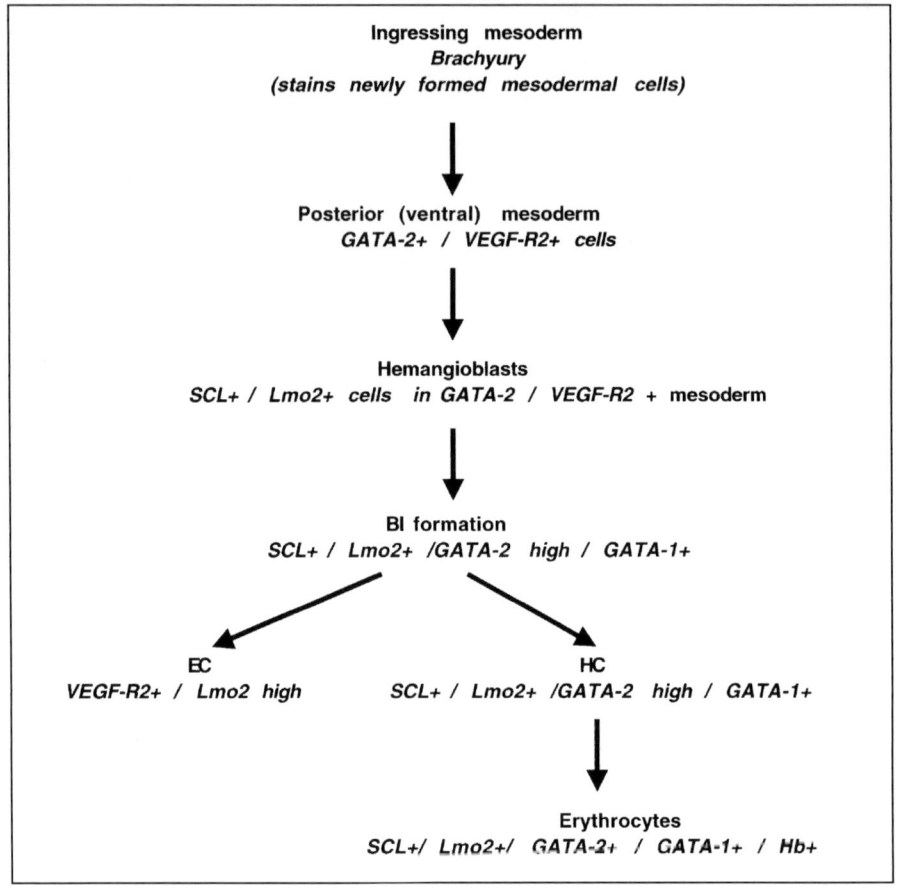

Figure 3. Pathway of YS erythropoiesis. Abbreviations: HB: hemoglobin; EC: endothelial cell; HC: hematopoietic cell; EryP: primitive erythrocyte.

here as vascular islands) that never develop hematopoiesis although this area was in continuity with the AO. The blood forming potential of this area is revealed two days later as the aortic region undergoes hematopoiesis. We have probed these early events by examining the patterns of several molecules likely to be involved in HC and EC emergence. The study included transcription factors which participate in multimeric complexes: *GATA-1, -2, SCL/tal-1* and *Lmo2*, (whose avian orthologue we have cloned) and *VEGF-R2*, a regulator of hematopoietic and endothelial commitment. *VEGF-R2* mRNA distribution has already been described for the early avian blastodisc,[63] but its protein distribution in the germ layers had not been investigated in detail previously. Several unexpected findings were revealed: as soon as gastrulation proceeds at Hamburger and Hamilton stages (HH, 1951) 3 and 4 (respectively 12-13h and 18h of incubation) GATA-2 and VEGF-R2 were detected the earliest in the whole ingressing mesoderm. Two distinct patterns were evident; one low associated to the ingressing mesoderm, one high, specific, restricted to mesodermal cells in direct contact with the endoderm (Fig. 1) although no morphologic segregation was visible in the mesodermal compartment. We interpret this pattern as an early commitment event resulting from endoderm/mesoderm interactions. From this time, SCL/tal-1 and Lmo2 expression appeared in groups of cells or single cells likely to be hemangioblasts (the putative common precursor for endothelial and hematopoietic cells in the YS). Thus YS hemangioblasts displayed SCL/Tal-1, Lmo2 and shared expression of GATA-2 and VEGF-R2 with the rest of the mesoderm. This is followed by the upregulation of GATA-2 in hemangioblastic aggregates (Fig. 2). These gene expressions were preferentially found in cells located in the ventral aspect of the mesoderm and were accompanied by the appearance of morphologically distinct structures. At later stages, these structures became more conspicuous accompanied by the expression of SCL/tal1, Lmo2 and the appearance of GATA-1 (not shown). Interestingly EC retained VEGF-R2 expression and upregulated Lmo2 that appeared as an early marker for EC commitment (not shown). Noticeably, hemangioblastic aggregates appeared completely segregated from the overlying mesoderm and in close contact with the endoderm (Fig. 2). This segregation is also conspicuous in the embryonic area proper where the forerunners of the aorta (that will be designed thereafter as the vascular islands) appeared closely associated to the AP endoderm (not shown). In the next steps, BI mature and release HC. On the basis of these patterns deduced from in situ studies, we were able to propose the following pathway for YS erythropoiesis (Fig. 3).

Role of the Endoderm

We have seen that BI and vascular islands differentiate in close association with the endoderm and even appear segregated from the splanchnopleural mesoderm from which they originate. Extraembryonic and embryonic mesoderm provide different derivatives of the blood-forming system. The extraembryonic mesoderm gives rise to YS BI, whereas the embryonic mesoderm gives vascular islands that never undergo hematopoiesis. This difference is likely to result from different instructive capacities of either the mesoderm proper or the subjacent endoderm. Interestingly, AO and AP endoderms have different embryonic origins; AO endoderm originates from the endoblast deposited during segmentation and early phases of gastrulation whereas AP endoderm, that will constitute the future gut endoderm, is deposited at later steps of gastrulation from cells ingressing through the primitive streak replacing the hypoblast that is pushed away to the edges of the AP. These morphogenetic movements give rise to a specific endoderm patterning that may constitute a basis for different instructive capacities and might explain differences in the developmental programs in the AO and AP blood-forming systems.

Contact with endoderm has long been thought to play a critical role in the specification of the blood-forming system. As early as in the middle 60s, Wilt and coworkers demonstrated that the absence of endoderm abolishes the differentiation of EC.[60] In the next decades, classical tissue recombination studies suggested that blood system formation requires diffusible signals from the adjacent endoderm.[6,43,51,62] Various growth factors may be responsible for this endodermal effect (notably bFGF and VEGF). For instance, bFGF can substitute for the chick endoderm,[26] VEGF is present in the mouse,[7,22,42] avian,[1,25] Xenopus[12] and zebrafish[58] YS endoderm. In mouse VEGF Knock Out (KO), the loss of a single allele is sufficient to impair both HC and EC development[10,24] demonstrating that gene dosage is critical for these processes. Moreover, VEGF-R2 KO produces the same phenotype.[56,57] It is known that VEGF mediates, multiplication, migration and chemotaxis of EC. Taking into account that a specific population of VEGF-R2+ and GATA-2+ cells is detected in close association with the endoderm, VEGF could attract VEGF-R2+ cells to the endoderm creating a layer of cells receiving directly further instructive endodermal signals. Recently Indian hedgehog was demonstrated to have a key role in yolk sac BI specification in the mouse.[23] Additional work will be required to identify the signal(s) responsible for the different AO/AP mesoderm instructions.

Hemangioblast and the Hemogenic Endothelium

When YS BI form, HC and EC emerge at the same time and in close association. In the aorta, HC are intimately associated to EC. The association between these two cells types has prompted earlier embryologists to assume the existence of a putative ancestral progenitor between EC and HC called the hemangioblast. The existence of this common progenitor has been assumed at that time for the YS where blood formation is conspicuous and available to direct observation. This hypothesis has been reinforced by the fact that 1° both cell types shared several markers and 2° several gene mutations and deletions both in zebrafish and mouse embryos affect both cell types. Whereas the existence of this elusive cell type has been demonstrated in ES cell cultures,[11] such a common ancestor has not been shown yet in vivo. The avian data that we will consider here even argues against such an existence. The avian homologue of VEGF-R2, Quek-1 was shown to be expressed by the mesodermal progenitors giving rise to YS mesoderm. Aiming to identify the hemangioblast, Eichmannn et al[65] sorted out Quek-1+ cells from the very early blastodisc and submitted the positive cell population to a clonal differentiation assay. In the absence of growth factors, VEGF-R2+ cells gave rise to hematopoietic colonies. When VEGF was added, endothelial colonies differentiated while the number of hematopoietic colonies decreased substantially. Mixed colonies never developed precluding the demonstration of a common precursor. Moreover, recent mapping experiments in the mouse embryo even suggest that HC and EC are segregated separately during the gastrulation process; HC pass through the PS first followed by the EC population.[33]

If filiation between HC and EC is not solved yet for the YS BI, generation of HC in the aortic region has been analysed in much more detail. In the E3 aorta, hematopoietic cells appear to delaminate from the aortic floor endothelium. Similar groups of cells have been described in the same location in all vertebrate species investigated. We first analysed the characteristics of the cells lining the aortic lumen at the time of hematopoietic emergence using double staining with antibodies directed against CD45, the pan-leukocyte antigen and VEGF-R2 specific for EC at that time.[28] Before cluster emergence, aortic EC were entirely positive for VEGF-R2. At the time of cluster differentiation, VEGF-R2 was downgraded and CD45 upgraded in EC of the ventral endothelium. As hematopoietic cells bulge in the aortic lumen, rounded cells become all CD45+.[28]

In order to obtain a dynamic approach of the developmental relationships between cluster cells and the aortic endothelium, we used two vital tracers. The first, acetylated low density lipoprotein (AcLDL) coupled to a fluorescent lipophilic marker (DiI) was inoculated in the E2 vascular tree one day before the emergence of intraaortic clusters. This marker has two major advantages: 1° it is specifically endocytosed by EC and macrophages (which are not present in the embryo at that stage) via a specific receptor whereas the other cell types never uptake the molecule. 2° the compound displays a short half-life that precludes its presence in the circulation for a long period of time if not endocytosed. As early as two hours after inoculation, the whole vascular tree was labelled. The aortae, still paired at that time, was entirely lined by AcLDL+ EC. Twenty four hours after inoculation, the intraaortic clusters and the rest of the EC were entirely AcLDL+. Moreover AcLDL+ cells were found in the mesenchyme ventral to the aorta which suggests that a subset of HC ingresses into this region.[28] It was thus clear that the intraaortic clusters were derived from cells that previously had an endothelial phenotype at the time of inoculation and underwent transdifferentiation during cluster emergence. Thus, clusters were generated through an EC intermediate located in the aortic floor. Of importance is the recent demonstration in mouse and human embryos that aortic EC can also give rise to HC.[19,49]

Contrary to the intraaortic clusters, paraaortic foci develop in a mesenchyme ventral to the aorta and thus appear to have no relationship with the aortic endothelium. Experimental analysis has shown that these cells actually contained HSC. Was there a developmental

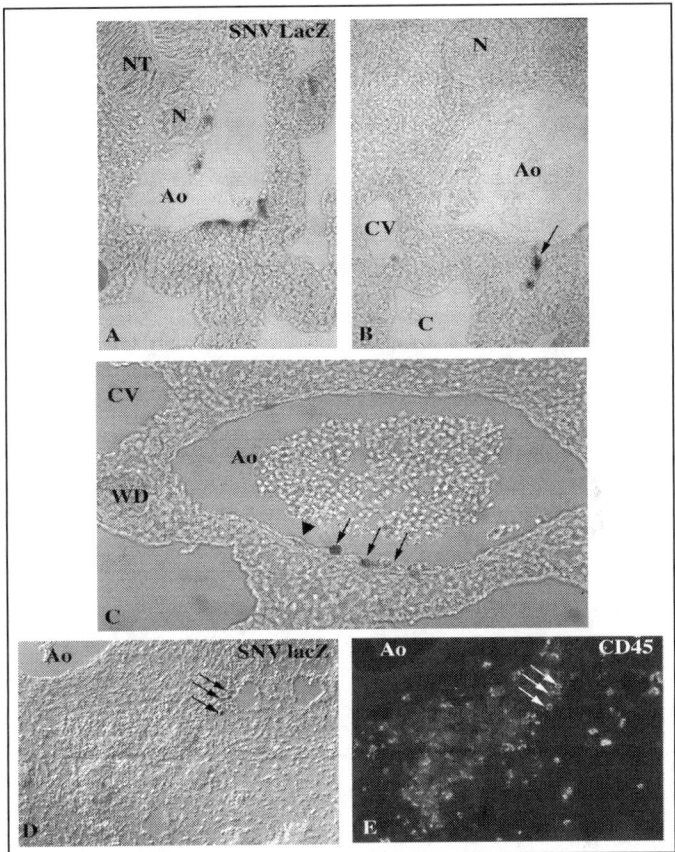

Figure 4. Tracing the hemogenic endothelium with nonreplicative retroviral vectors during formation of the intraaortic clusters. Embryos have been inoculated at E2 with the vectors. A, B, C, D, LacZ staining and Nomarski's interference contrast. E) CD45 immunohistochemistry. A)Early day three. One day after viral inoculation. EC of the aortic floor and roof display lacZ expression. B) 36h after retroviral inoculation. Cell ingressing from the aortic floor with a long filopode still attached close to the aortic lumen (arrow). C) 36h after inoculation, mid trunk level. Clusters (arrows) and EC (arrowhead) display lacZ expression. D) E6, four days after inoculation. LacZ+ cells are detected in the ventral aortic mesenchyme (black arrows). E) CD45 staining. LacZ+ cells display a high CD45 expression (white arrows). C: coelom; CV: cardinal vein; WD: wolffian duct.

relationship between the two aspects of aortic hemopoiesis or did they belong to two independent generations of HSC? This question is of importance since the adult is no longer capable of producing HSC. In other terms are HSC obviously generated through an endothelium intermediate during early development or could they be generated de novo from a mesenchyme. To approach this question we have used nonreplicative lacZ-bearing retroviral vectors inoculated in the same conditions as for the AcLDL-DiI. The interest of using this system is two folds 1° clones can be identified because integration occurs in a few dispersed cells of the endothelial tree. 2° the reporter gene is stably expressed in the progeny of infected cells allowing to follow tagged cells for long periods of time. The conclusions were as follows: At E3 the labelled clones comprised either EC and HC but never both. Labelled cells were observed to ingress into the mesentery ventral to the aortic floor. At E7, the paraaortic foci contained numerous

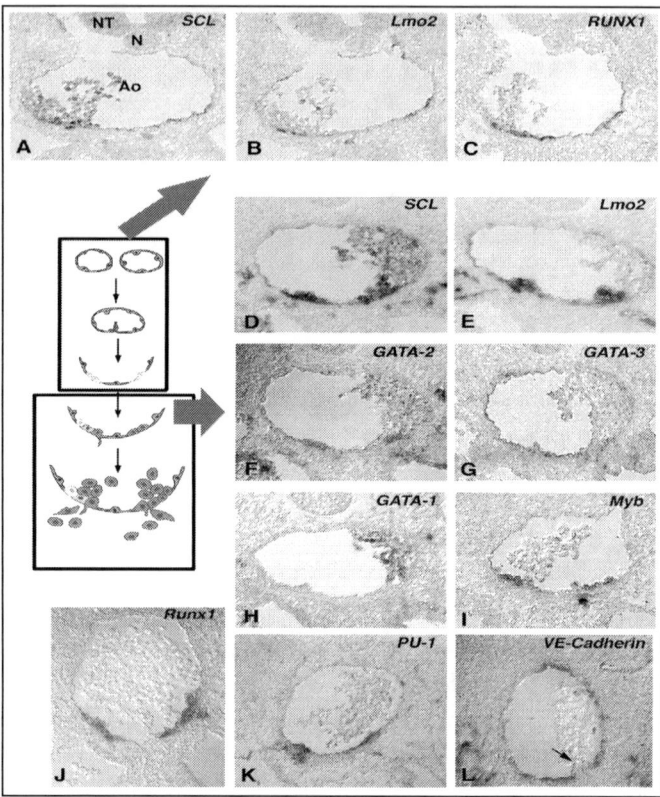

Figure 5. Gene activities during formation of the intraaortic clusters. The drawing in the left side schematises the different steps in the formation of the clusters. The frames point out the two major aspects of cluster formation: hemogenic endothelium without any morphological sign of hematopoiesis and full cluster formation. A, B, C) Beginning of the third day of development when intraaortic clusters have not appeared yet. Adjacent sections. D to L) Middle of the third day when intraaortic clusters are fully developed. A) SCL/Tal-1 is expressed by all aortic EC but appears strengthened in the aortic floor. Note the presence of SCL/Tal-1+ circulating cells. B) Lmo2 stains all aortic EC. A few circulating cells are Lmo2+. C) Runx1 expression is restricted to EC of the aortic floor. D) SCL/Tal-1 is now vividly expressed by the clusters and expressed at low levels by aortic EC. E) Lmo2 displays a similar signal. In addition non-aortic EC display a conspicuous Lmo2 expression. F) GATA-2 expression is restricted to the clusters. A weak or no signal is detected in EC. G) GATA-3 stains EC but is also detected, although at low levels in the clusters. H) GATA-1 is absent. I) Myb is vividly expressed in cluster cells as well as in ingressing cells in the mesenchyme. J) Runx1 is specifically expressed by cluster cells and ingressing cells. K) PU-1, a target gene for Runx1 is detected into the same cells as those expressing Runx1. L) VE-Cadherin stains vascular EC but is absent from the clusters (arrow). Ao: aorta; N: notochord; NT: neural tube.

lacZ+ cells (Fig. 4).[29] This demonstrated that paraortic foci are seeded by progenitors derived from the E3 aortic floor thus demonstrating the existence of a unique event of HSC emergence from the aortic endothelium.

Using molecular markers, several of which we have cloned, we have been able to identify several steps in the transition from EC to HC. A first, when the aortae are still paired and no sign of hematopoiesis is visible. The whole aortic EC express molecules of a typical endothelium (the surface molecules VEGF-R2, VE-Cadherin, and the low expression of transcription factors Lmo2, SCL/tal-1). A second that occurs when the aortae are still paired or when they undergo fusion. Runx1 and GATA-2 expression appeared and SCL/Tal-1 and Lmo2, became upregulated in the ventral aspect of the aorta that also retain expression of the EC-specific genes at that time. No morphological sign of HC emergence is visible yet. A third when fusion has occurred; characterised by the thickening of ventral EC accompanied by the increased expression of HC-specific genes and the decrease of EC's (downgrading of VEGF-R2, VE-Cadherin; upgrading of Runx1, SCL/tal-1, Lmo2, GATA-2, GATA-3). A fourth one in which clusters are conspicuous where expressions of EC and HC-specific genes are dissociated (Fig. 5). Runx1, the avian orthologue of which we have cloned, is the earliest gene patterned to the ventral region of the aorta. It belongs to the family of transcription factors with a runt domain. Inactivation of this gene in the mouse embryo precludes the emergence of the definitive hematopoiesis.[59] Using a knock in approach, Runx1 was shown to be specifically expressed in the ventro-lateral aortic EC at the time, but even before cluster emergence.[48] It is to date the only gene harbouring such a pattern. In the avian embryo, Runx1 is expressed as early as E2.5 in EC of the ventral aortic floor. Expression begins when the aortae are still paired and persisted upon fusion of the vessels. At the time of hematopoietic cluster production, Runx1 becomes restricted to HC bulging into the lumen and ingressing into the mesentery. Interestingly PU-1, a target gene for Runx1, expressed in precursors of the myeloid and lymphoid lineages is detected in intraaortic clusters (Fig. 5).

Dual Origin of Aortic ECs

What makes the aorta regionalized in its capacity to provide HSC? Indeed the ventral aspect of the vessel is endowed with hematopoietic capacities whereas the dorsal aspect never develops hematopoiesis. Several groups have shown that somites produce angioblasts.[3,47,55] Orthotopic exchanges of somites from quail into chicken embryos were done involving the last-formed somites which from cells have not emigrated yet. QH1+ cells localized into the host endothelia of the body wall, kidney, aortic roof and sides. In contrast these cells never invaded the visceral organs or the aortic floor. When chick host somites were replaced by quail splanchnopleural mesoderm, the same derivatives were colonized but, in addition, QH1+ cells localized into the visceral organs and aortic floor where some displayed a hematopoietic phenotype.[52] Thus two distinct lineages become committed in the embryo proper, a dorsal, angioblastic, lineage originating from the somites which have never been in contact with the endoderm, endowed uniquely with endothelial properties and a ventral, hemangioblastic, splanchnopleural, derivative which has been in close contact with the endoderm during its history, participating to the building of the ventral aspect of the aorta.[52] The aorta has thus a dual origin responsible for the restriction of HC emergence in the aortic floor. Finally the body wall mesoderm is vascularized by somitic EC.

Manipulating the Angiopoietic/Hemangiopoietic Commitment

This last finding suggests that the aorta is composed of two pools of EC that are differentially influenced according to their location in the embryo: a ventral pool that develop in close contact with the endoderm and a dorsal pool influenced by the ectoderm. These two influences were probed on different subsets of mesoderm using quail-into-chick grafts followed by analysis with the QH1 Mab. The promise was to know whether it was possible to switch the mesodermal fate from dorsal to ventral or inversely. The experimental scheme was as followed. Firstly, explants of mesoderm were treated in vitro for 12 hours either by contact with ectoderm, endoderm or by various growth factors. Secondly the explants were grafted into a chick host coelom and localization of QH1+ cells was analysed. When treated with endoderm, VEGF, bFGF and TGFβ1, somatic or somatopleural mesoderm were capable of generating QH1+ cells that invade ventral structures i.e., internal organs, and ventral floor of the aorta.

Ectoderm, EGF and TGFα were capable of deleting the ventral potential of splanchnopleural mesoderm restricting the migration of QH1+ cells to dorsal structures.[51] These conclusions agree with and extend that of earlier studies suggesting a positive influence of endoderm[31,32,60] and a negative influence of ectoderm on hematopoiesis.[31,32] Further work is required to know how and when these molecules, and maybe others, instruct mesodermal precursors.

The Allantois, a Later Source of HSC Colonizing the Bone Marrow

In birds, the timing of intraaortic clusters and paraaortic foci is compatible with the seeding of the thymus and bursa of Fabricius. Cells of the paraaortic foci have been shown to have B and T lymphoid potentials.[34] However, seeding of the bone marrow begins by E10.5 in the chick embryo and continues until birth. At this time, the activity of the paraaortic foci has ceased. The possibility that cells from the paraaortic foci remained in an undifferentiated state somewhere into the embryo appeared unlikely. We thus have looked for another progenitor-producing site and identified the allantois. This organ has the appropriate tissue make-up to produce hemopoiesis i.e., endoderm associated to mesoderm. This appendage was shown to be involved in gas exchange, excretion product accumulation, shell Ca++ resorbtion and bone construction. By using India ink or AcLDL-Di microangiographies, we first established that the allantois became vascularized between 75-80 hours of incubation. It displayed conspicuous red cells even before this period indicating that hematopoiesis occurred independently from the rest of the embryo in this site.[8] A full cellular and molecular hematopoietic program, characterized by the expression of SCL/tal-1, Lmo2 GATA-2 and GATA-1 and the emergence of a second generation of red cells occurred in the allantoic mesoderm. This is accompanied by the expression of GATA-2 and GATA-3 in the associated endoderm.[9] The allantoic rudiment was dissected out prior to vascularisation and grafted heterotopically into the coelom of E3 chicks. The bone marrow of E13 chicks was found to harbor as significant amount of QH1+ cells, up to 8% of the total bone marrow cells. Analysis on sections also revealed that QH1+ cells were not only hematopoietic but also endothelial in nature.[8] Although seeding of distinct progenitors was already shown, the possibility of the seeding a common progenitor cannot be excluded. Recent work in the mouse embryo indicates that the fetal part of the placenta that derived from the allantois is a source of hematopoietic progenitors (Alvarez-Silva, Belo-Diabangouaya, Salaun, Dieterlen-Lièvre, submitted).

Conclusions

The demonstration of an intraembryonic source of hematopoietic progenitors has paved the way for investigations in other vertebrates. It is now strongly established that the YS gives only transient progenitors whereas the embryo proper gives rise to HSC. How these programs are established is not known but they are likely influenced by differences in the microenvironments. What makes YS hematopoietic cells to readily differentiate whereas the aortic cells retain multipotential capacities? What molecular program triggers aortic EC to give rise to HSC and what are the roles of several key transcription factors in this aortic hematopoietic emergence (Runx1, CBFβ, SCL/tal-1) will be the next challenges. There is no doubt that the avian embryo will be one of the major actors of this play.

Acknowledgements

We thank Dr. F. Dieterlen for constant support throughout this work. This study was supported by the Centre National pour la Recherche Scientifique, by grants from the Association pour la Recherche Contre le Cancer N°5131, from the Fondation pour la Recherche Médicale and by an MERT ACI N° 22-2002-296. KM is a recipient of ARC and FRM predoctoral fellowships. KB is recipient of a French MERT fellowship N° 00391.

References

1. Aitkenhead M, Christ B, Eichmann A et al. Paracrine and autocrine regulation of vascular endothelial growth factor during tissue differentiation in the quail. Dev Dyn 1998; 212:1-13.
2. Alliot F, Godin I. Pessac B. Microglia derive from progenitors, originating from the yolk sac, and which proliferate in the brain. Brain Res Dev Brain Res 1999; 117:145-152.
3. Ambler CA, Nowicki JL, Burke AC et al. Assembly of trunk and limb blood vessels involves extensive migration and vasculogenesis of somite-derived angioblasts. Dev Biol 2001; 234:352-364.
4. Beaupain D, Martin C, Dieterlen-Lievre F. Are developmental hemoglobin changes related to the origin of stem cells and site of erythropoiesis? Blood 1979; 53:212-225.
5. Beaupain D, Martin C, Dieterlen-Lievre F. Origin and evolution of hemopoietic stem cells in avian embryo. In: Stammatoyanopoulos G, Nienhuis A, eds. Hemoglobins in development and differentiation. New York: Alan R Liss, Inc, 1979:161-169.
6. Belaoussoff M, Farrington SM, Baron MH. Hematopoietic induction and respecification of A-P identity by visceral endoderm signaling in the mouse embryo. Development 1998; 125:5009-5018.
7. Breier G, Clauss M, Risau W. Coordinate expression of vascular endothelial growth factor receptor-1 (flt-1) and its ligand suggests a paracrine regulation of murine vascular development. Dev Dyn 1995; 204:228-239.
8. Caprioli A, Jaffredo T, Gautier R et al. Blood-borne seeding by hematopoietic and endothelial precursors from the allantois. Proc Natl Acad Sci USA 1998; 95:1641-1646.
9. Caprioli A, Minko K, Drevon C et al. Hemangioblast commitment in the avian allantois: cellular and molecular aspects. Dev Biol 2001; 238:64-78.
10. Carmeliet P, Ferreira V, Breier G et al. Abnormal blood vessel development and lethality in embryos lacking a single VEGF allele. Nature 1996; 380:435-439.
11. Choi K, Kennedy M, Kazarov A et al. A common precursor for hematopoietic and endothelial cells. Development 1998; 125:725-732.
12. Cleaver O, Krieg PA. VEGF mediates angioblast migration during development of the dorsal aorta in Xenopus. Development 1998; 125:3905-3914.
13. Cooper MD, Chen CH, Bucy RP et al. Avian T cell ontogeny. Adv Immunol 1991; 30:87.
14. Cormier F. Avian pluripotent haemopoietic progenitor cells: detection and enrichment from the para-aortic region of the early embryo. J Cell Sci 1993; 105:661-666.
15. Cormier F, de Paz P, Dieterlen-Lievre F. In vitro detection of cells with monocytic potentiality in the wall of the chick embryo aorta. Dev Biol 1986; 118:167-175.
16. Cormier F, Dieterlen-Lievre F. The wall of the chick embryo aorta harbours M-CFC, G-CFC, GM-CFC and BFU-E. Development 1988; 102:279-285.
17. Cuadros MA, Coltey P, Carmen Nieto M et al. Demonstration of a phagocytic cell system belonging to the hemopoietic lineage and originating from the yolk sac in the early avian embryo. Development 1992; 115:157-168.
18. Cuadros MA, Navascues J. Early origin and colonization of the developing central nervous system by microglial precursors. Prog Brain Res 2001; 132:51-59.
19. de Bruijn MF, Ma X, Robin C et al. Hematopoietic stem cells localize to the endothelial cell layer in the midgestation mouse aorta. Immunity 2002; 16:673-683.
20. Dieterlen-Lievre F. On the origin of haemopoietic stem cells in the avian embryo: an experimental approach. J Embryol Exp Morphol 1975; 33:607-619.
21. Dieterlen-Lievre F, Martin C. Diffuse intraembryonic hemopoiesis in normal and chimeric avian development. Dev Biol 1981; 88:180-191.
22. Dumont DJ, Fong GH, Puri MC et al. Vascularization of the mouse embryo: a study of flk-1, tek, tie, and vascular endothelial growth factor expression during development. Dev Dyn 1995; 203:80-92.
23. Dyer MA, Farrington SM, Mohn D et al. Indian hedgehog activates hematopoiesis and vasculogenesis and can respecify prospective neurectodermal cell fate in the mouse embryo. Development 2001; 128:1717-1730.
24. Ferrara N, Carver-Moore K, Chen H et al. Heterozygous embryonic lethality induced by targeted inactivation of the VEGF gene. Nature 1996; 380:439-442.

25. Flamme I, Breier G, Risau W. Vascular endothelial growth factor (VEGF) and VEGF receptor 2 (flk-1) are expressed during vasculogenesis and vascular differentiation in the quail embryo. Dev Biol 1995; 169:699-712.
26. Gordon-Thomson C, Fabian BC. Hypoblastic tissue and fibroblast growth factor induce blood tissue (haemoglobin) in the early chick embryo. Development 1994; 120:3571-3579.
27. Jaffredo T, Gautier R, Brajeul V et al. Tracing the progeny of the aortic hemangioblast in the avian embryo. Dev Biol 2000; 224:204-214.
28. Jaffredo T, Gautier R, Eichmann A et al. Intraaortic hemopoietic cells are derived from endothelial cells during ontogeny. Development 1998; 125:4575-4583.
29. Jotereau F, Houssaint E. Experimental studies on the migration and differentiation of primary lymphoid stem cells in the avian embryo. In: Solomon JB, Horton JD, eds. Developmental Immunobiology. Amsterdam: Elsevier North Holland Biomedical Press, 1977.
30. Jotereau FV, Le Douarin NM. Demonstration of a cyclic renewal of the lymphocyte precursor cells in the quail thymus during embryonic and perinatal life. J Immunol 1982; 129:1869-1877.
31. Kessel J, Fabian B. Inhibitory and stimulatory influences on mesodermal erythropoiesis in the early chick blastoderm. Development 1987; 101:45-49.
32. Kessel J, Fabian BC. The pluripotency of the extraembryonic mesodermal cells of the early chick blastoderm: effects of the AP and AOV environments. Dev Biol 1986; 116:319-327.
33. Kinder SJ, Tsang TE, Quinlan GA et al. The orderly allocation of mesodermal cells to the extraembryonic structures and the anteroposterior axis during gastrulation of the mouse embryo. Development 1999; 126:4691-4701.
34. Lassila O, Eskola J, Toivanen P. Prebursal stem cells in the intraembryonic mesenchyme of the chick embryo at 7 days of incubation. J Immunol 1979; 123:2091-2094.
35. Lassila O, Eskola J, Toivanen P et al. The origin of lymphoid stem cells studied in chick yold sac-embryo chimaeras. Nature 1978; 272:353-354.
36. Lassila O, Martin C, Dieterlen-Lievre F et al. Is the yolk sac the primary origin of lymphoid stem cells? Transplant Proc 1979; 11:1085-1088.
37. Lassila O, Martin C, Toivanen P et al. Erythropoiesis and lymphopoiesis in the chick yolk-sac-embryo chimeras: contribution of yolk sac and intraembryonic stem cells. Blood 1982; 59:377-381.
38. Le Douarin N, Jotereau F, Houssaint E et al. Primary lymphoid organ ontogeny in birds. In: Le Douarin NLM, McLaren A, eds. Chimeras in developmental biology. London: Academic Prss, 1984:179.
39. Le Douarin NM, Dieterlen-Lievre F, Oliver PD. Ontogeny of primary lymphoid organs and lymphoid stem cells. Am J Anat 1984; 170:261-299.
40. Martin C, Beaupain D, Dieterlen-Lievre F. Developmental relationships between vitelline and intra-embryonic haemopoiesis studied in avian 'yolk sac chimaeras'. Cell Differ 1978; 7:115-130.
41. Martin C, Lassila O, Nurmi T et al. Intraembryonic origin of lymphoid stem cells in the chicken: studies with sex chromosome and IgG allotype markers in histocompatible yolk sac-embryo chimaeras. Scand J Immunol 1979; 10:333-338.
42. Miquerol L, Gertsenstein M, Harpal K et al. Multiple developmental roles of VEGF suggested by a LacZ-tagged allele. Dev Biol 1999; 212:307-322.
43. Miura Y, Wilt FH. Tissue interaction and the formation of the first erythroblasts of the chick embryo. Dev Biol 1969; 19:201-211.
44. Moore MAT, Owen JJT. Chromosome marker studies in the development of the hematopoietic system in the chick embryo. Nature 1965; 208:956.
45. Moore MAT, Owen JJT. Experimental studies on the development of the thymus. J Exp Med 1967; 126:715.
46. Moore MAT, Owen JJT. Stem cell migration in developing myeloid and lyphoid systems. Lancet 1967; 2:658.
47. Noden DM. Embryonic origins and assembly of blood vessels. Am Rev Respir Dis 1989; 140:1097-1103.
48. North T, Gu TL, Stacy T et al. Cbfa2 is required for the formation of intra-aortic hematopoietic clusters. Development 1999; 126:2563-2575.

49. Oberlin E, Tavian M, Blazsek I et al. Blood-forming potential of vascular endothelium in the human embryo. Development 2002; 129:4147-4157.

50. Pardanaud L, Altmann C, Kitos P et al. Vasculogenesis in the early quail blastodisc as studied with a monoclonal antibody recognizing endothelial cells. Development 1987; 100:339-349.

51. Pardanaud L, Dieterlen-Lievre Manipulation of the angiopoietic/hemangiopoietic commitment in the avian embryo. Development 1999; 126:617-627.

52. Pardanaud L, Luton D, Prigent M et al. Two distinct endothelial lineages in ontogeny, one of them related to hemopoiesis. Development 1996; 122:1363-1371.

53. Peault BM, Thiery JP, Le Douarin NM. Surface marker for hemopoietic and endothelial cell lineages in quail that is defined by a monoclonal antibody. Proc Natl Acad Sci USA 1983; 80:2976-80.

54. Schoenwolf GC, Garcia-Martinez V, Dias MS. Mesoderm movement and fate during avian gastrulation and neurulation. Dev Dyn 1992; 193:235-248.

55. Schramm C, Solursh M. The formation of premuscle masses during chick wing bud development. Anat Embryol (Berl) 1990; 182:235-247.

56. Shalaby F, Ho J, Stanford WL et al. A requirement for Flk1 in primitive and definitive hematopoiesis and vasculogenesis. Cell 1997; 89:981-990.

57. Shalaby F, Rossant J, Yamaguchi TP et al. Failure of blood-island formation and vasculogenesis in Flk-1-deficient mice. Nature 1995; 376:62-66.

58. Thompson MA, Ransom DG, Pratt SJ et al. The cloche and spadetail genes differentially affect hematopoiesis and vasculogenesis. Dev Biol 1998; 197:248-269.

59. Wang Q, Stacy T, Binder M et al. Disruption of the Cbfa2 gene causes necrosis and hemorrhaging in the central nervous system and blocks definitive hematopoiesis. Proc Natl Acad Sci USA 1996; 93:3444-3449.

60. Wilt FH. Erythropoiesis in the chick embryo: The role of the endoderm. Science 1965; 147:1588-1590.

61. Wilting J, Brand-Saberi B, Huang R et al. Angiogenic potential of the avian somite. Dev Dyn 1995; 202:165-171.

62. Zagris N. Hypoblast induction of multiple areas vasculosae, and stabilization of the area opaca vasculosa in young chick blastoderm. J Embryol Exp Morphol 1982; 68:115-126.

63. Eichmann A, Marcelle C, Bréant C et al. Two molecules related to the VEGF receptor are expressed in early endothelial cells during avian embryonic development. Mech Dev 1993; 42:33-48.

64. Hamburger V, Hamilton HL. A series of normal stages in the development of the chick embryo. Morphol 1951; 88:49-92.

65. Eichmann A, Corbel C, Nataf V et al. Ligand-dependent development of the endothelial and hematopoietic lineages from embryonic mesodermal cells expressing vascular endothelial growth factor receptor 2. Proc Natl Acad Sci USA 1997; 94:5141-5146.

CHAPTER 4

Avian Lymphopoiesis and Transcriptional Control of Hematopoietic Stem Cell Differentiation

Jussi Liippo and Olli Lassila

Abstract

Hematopoiesis starts in the yolk sac blood islands and in a region of dorsal aorta in the early embryo. Interest in the emergence of the hematopoietic stem cells in the embryo proper has increased in last years. In an avian model the primitive erythropoiesis derived from extraembryonic yolk sac is short lived and the evidence that adult hematopoiesis arises from intraembryonic sources has been extended from birds to mammals. Definitive lymphohematopoiesis initiates in the ventral wall of the dorsal aorta in all the vertebrate species.

Second major issue in early hematopoiesis is the existence of hemangioblast, a bipotential cell, able to differentiate into a hematopoietic stem cell or a vascular endothelial cell. It has become clear that the emergence of hematopoietic stem cells and hemangioblasts is tightly linked. The rapid commitment of hematopoietic and endothelial lineages in the developing embryo indicates that the distinction between mesoderm and hemangioblast maybe difficult to define. It is, however, possible that the hemangioblast stage is very short lived.

Significant new insights have been gained in vertebrate species in our understanding of lineage determination and commitment of hematopoietic stem cells into various lineages. It has been shown that lineage specific transcription factors have essential role in fate decisions. Recent work has shown that these factors function as activators or repressors showing dynamic balance determining the phenotype of the cell. Some of these factors are also actively repressing the alternate lineage gene expression. The genetic programmes controlled by transcriptional regulators seem well conserved between vertebrate species.

As the intraembryonic hemogenic sites start to seed the lymphoid organ rudiments, developmental programs of the homing progenitors are gradually restricted towards downstream lineages. Eventually, thymus, bone marrow, spleen and the avian bursa of Fabricius serve as permissive environments for further maturation, proliferative expansion and selection processes. During these complex events multiple both cell-intrinsic and -extrinsic mediators critically influence the emergence of lymphohemopoietic progeny.

Hematopoietic Stem Cell Development, edited by Isabelle Godin and Ana Cumano.
©2006 Eurekah.com and Kluwer Academic / Plenum Publishers.

Hemangioblasts

Following fertilization the totipotent embryonic stem (ES) cells of vertebrate embryos divide rapidly to produce multipotent stem cells that in turn reside at the top of various developmental cascades. These processes involving a series of developmental choices are naturally subjected to a tightly orchestrated and controlled usage of intranuclear genetic potential. One of the most thoroughly analysed developmental pathways is the hemangiopoietic system (Fig. 1). Although not completely established, several lines of evidence have recently suggested that the vascular and hematopoietic systems share a common developmental progeny.[1-3] Anatomically, both the hematopoietic and endothelial stem cells colocalize to the ventral para-aortic splanchnopleural mesoderm and the avian allantois.[4-6] The hypothetical bipotent progenitor, however, the hemangioblast (Fig. 1), has not been identified at a clonal level so far. Still, both the hematopoietic and endothelial cells have many common antigenic features.[7-9] More importantly, gene targeting and other experiments have documented the central role of transcription regulators (Table 1), the stem cell leukemia factor (SCL/TAL-1) and acute myeloid leukemia factor 1 (AML1) as well as growth factor receptors, i.e., VEGF-R2/Flk-1, TGF-β and Tie-2/Tek for the proper development of both systems.[7,10,11]

SCL-deficient mice fail to survive due to severe defects in blood formation and the corresponding mutant ES cells are unable to contribute to any hematopoietic lineages (Table 1).[12] Later, however, it has become clear through transgenic rescue experiments that SCL has also a significant contribution to the vascular development and, specifically, to the angiogenic remodeling and branching of the preexisting capillary plexus (Fig. 1). Accordingly, the zebrafish SCL ortholog plays a role in the development of hemangiopoietic cells from the embryonic lateral mesoderm.[13] Other experiments using the zebrafish *cloche* mutants have in addition demonstrated that the hemangiopoietic defect present in these mutants was due to abolished SCL expression. Moreover, these studies were able to show that SCL acts upstream of at least Flk-1, Tie-1 and GATA-1, as their expression together with both the hematopoietic and vascular defects were rescued via forced SCL expression. In conclusion, it has been suggested that SCL could trigger the emergence of vasculohemopoietic progeny from the ventral mesoderm in a similar way as other related bHLH-factors, like MyoD, control the onset of e.g., muscular development.[14] More specifically, AML1 (Cbfa2) that is required for definitive hematopoiesis is transiently expressed in the endothelial cells at sites of embryonic HSC emergence (Table 1, Fig. 1).[15] These studies have suggested that AML1 could mark adoption of hematopoietic fate from the presumptive hemangiopoietic precursors present in the early embryonic mesodermal sites. Recently, it has also been suggested that the angiogenic defects seen in *AML1*-deficient animals are due to the absence HSCs, which suggests a novel primary role for HSCs as promoters of angiogenic events in early embryos.[11]

Transcription Factors and Early Hematopoiesis

Following the delineation of vasculogenic and hemogenic progenies the intraembryonic aorta-associated sites function as the origin for the definitive hematopoietic series.[16-19] As the early yolk sac blood island-residing stem cells account only for a transitory phase of primitive erythroid development,[20] the intraembryonic ventral mesodermal foci harbor the true hematopoietic stem cells. As the hemangiopoietic development relies on a given set of molecular regulators, the hematopoietic system itself depends on another and less ubiquitous category of transcription controllers. One of the key factors in early hematopoiesis is the lim-domain containing Lmo2 (Rbtn2) that has, like SCL, been linked to leukemogenesis and is capable of forming heterodimers with SCL proteins in erythroid cells.[21] *Lmo2*-deficiency causes embryonic lethality via similar hematopoietic defects as seen in *SCL* mutants and mutant ES cells fail to generate any hematopoietic lineages in adult mice (Table 1, Fig. 1).

Table 1. Transcription factors of hematopoiesis

Factor	Family	Expression	Gene Targeting
SCL/Tal-1	bHLH	HSC	No primitive and definitive hematopoiesis, yolk sac angiogenesis defective
AML1 (Cbfa2)	CBF	HSC	No definitive hematopoiesis, angiogenic defects
Lmo2 (Rbtn2)	LIM domain	HSC	No primitive and definitive hematopoiesis
GATA-2	Zinc finger	HSC and precursors	HSC and precursors reduced
PU.1	Ets	Precursors and B cells	No myeloid and lymphoid cells
Ikaros	Zinc finger	HSC, T and B cells	1) Dominant negative homozygous: alymphoid 2) Dominant negative heterozygous: T cell hyperproliferation -> T cell leukemias and lymphomas 3) Null: lack of B, NK and fetal T cells, postnatal T cells undergo clonal expansion
Helios	Zinc finger	HSC and T cells	Unknown
E2A	bHLH	Constitutive	B lymphopoiesis arrested before Ig rearrangement
EBF	EBF	B lineage cells	B lymphopoiesis arrested before Ig rearrangement
Pax5	Paired box	B lineage cells	No B cells
Aiolos	Zinc finger	Lymphoid precursors and B cells	B cell hyperproliferation, disruption of B cell tolerance, B cell leukemias and lymphomas
GATA-3	Zinc finger	HSC, T cells and Th2	No T cells (block at the earliest double negative stage)
Tcf-1	HMG box	Thymocytes and T cells	T cell differentiation arrest due to proliferation defects in double negative thymocytes

Abbreviations: SCL: stem cell leukemia; bHLH: basic helix loop helix; HSC: hematopoietic stem cell; AML: acute myeloid leukemia; CBF: core binding factor; Lmo: LIM-only protein; EBF: early B cell factor; Tcf: T cell factor.

Another multilineage factor is the Ets family member PU.1 that is widely expressed in precursors of the lymphoid and myeloid lineages. Gene inactivation and retroviral transduction studies have concluded that PU.1 directs the entrance of HSCs into the lymphoid-myeloid pathways (Table 1, Fig. 1).[22] Specifically, high levels of PU.1 favor the development of macrophage progenitors, whereas lesser amounts induce the progenitors to adopt the B cell fate.[22] Whether these concentration-dependent choices function already at the HSC level or at bipotent myeloid-lymphoid precursor, are not fully understood. Common progenitors for myeloid and B cells have, however, been described in murine AGM (aorta-gonad-mesonephros) and fetal liver.[23] In addition, PU.1 is able to positively regulate its expression by binding to its own regulatory sites.[22]

The founding member of the GATA zinc finger family is GATA-1 which together with a cofactor FOG (Friend Of GATA-1) performs critical regulatory functions in the erythroid and

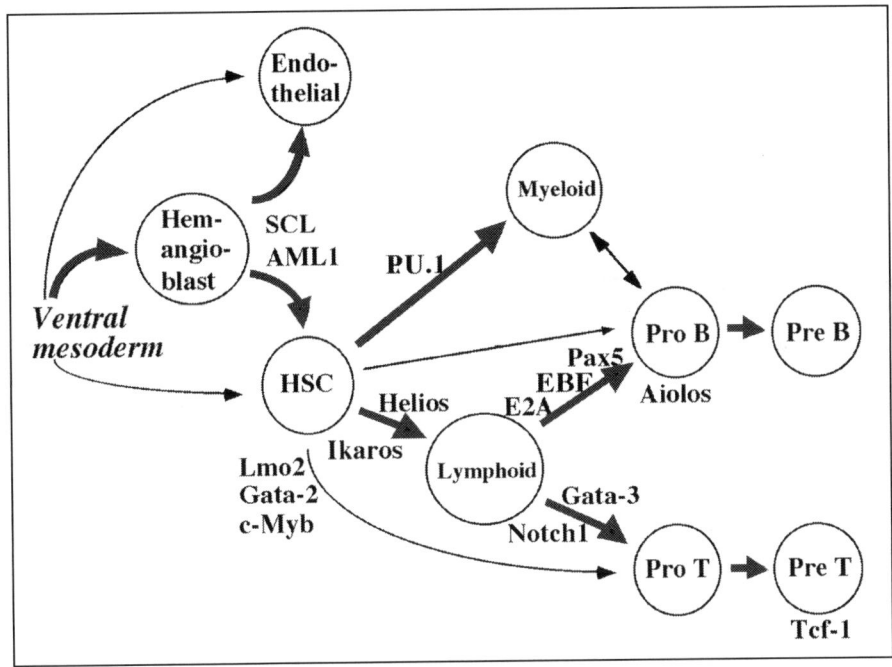

Figure 1. Central transcription regulators controlling the specification of the hemangiopoietic, hematopoietic, lymphoid and myeloid lineages. Embryonic ventral mesoderm is the primary origin for both the hematopoietic stem cells (HSC) and endothelial precursors. An elusive bipotent intermediate, the hemangioblast, has been proposed to represent a developmental intermediate. The illustrated transcription factors are placed according to their major sites of action. Thin arrows indicate alternate and more direct developmental routes.

megakaryocytic pathways.[24] Interestingly, the erythro-megakaryocytic lineage is rather unaffected in the *PU.1* mutant precursors indicating thus a distinct role for GATA-1 and PU.1 during the transition of HSC into myeloid and erythroid routes. In addition, forced GATA-1 expression redirects myeloid cells into erythroid or megakaryocytic lineages by repressing via protein-protein interactions the functional amount of PU.1.[25] Moreover, the antagonistic action of these factors is bidirectional, as also PU.1 is able to suppress GATA-1 in erythroid cells. Also during normal myeloid development GATA-1 expression is reduced, and analogously, PU.1 levels diminish upon the onset of erythroid differentiation. The second GATA-family member, GATA-2, has been suggested to control proliferative rate and growth factor responses of early hematopoietic precursors, as forced GATA-2 activity abolishes differentiation and enhances proliferation. Also in its absence all the definitive blood lineages are severely affected, which emphasizes the central role of GATA-2 at an early hematopoietic stage (Table 1, Fig. 1). The exact molecular mechanism of GATA-2 function, and the question of its own regulation, have still remained elusive. GATA-2 has not been shown to control its own expression by forming a positive autoregulatory loop like GATA-1 and PU.1. Interestingly, the Ikaros proteins, also members of this conserved zinc finger transcription factor family, have been demonstrated to play critical roles in proliferation control via chromatin remodeling and histone deacetylation.[26] Hence, as GATA-2 regulates proliferation of HSC, it would be intriguing to analyse, whether it has any interactions with the chromatin restructuring complexes controlling the accessibility of the target genes mediating this HSC proliferation.

Although GATA-1 has not either been shown to complex with the nucleosome modifying machinery, both GATA factors have been shown to interact with Lmo2 during early hematopoiesis. Indeed, it has been suggested that Lmo2 could function as an interface between GATA-2 and SCL in early HSCs and later, following commitment into the erythroid/megakaryocytic pathway, between SCL and GATA-1.[21] Putting this data together with the earlier gene inactivation and expression studies, a schematic picture could be drawn that puts first the gene causing the zebrafish *cloche* mutation upstream of SCL and Lmo2 that, in turn, are in front of GATA proteins in the developmental hierarchy of early HSCs (Fig. 1).

Adoption of the Lymphoid Fate

In vertebrate embryos the intraembryonic para-aortic mesoderm (avian) and the para-aortic splanchnopleura (mammals) and their derivative, the AGM region, are the primary origins for lymphoid progeny.[16-19,27] Already in the AGM and later in fetal liver of murine embryos, different progenitors with T-, B-, myeloid-, multilineage-, B/myeloid- and T/myeloid-capacities have been found.[23,28] In a similar way, progenitors with either B or T lymphoid characteristics are present in early avian embryos.[29,30] Moreover, recombination of the immunoglobulin genes occurs very early in the avian ontogeny.[31,32] Yet, although adult murine bone marrow has been suggested to contain hematopoietic cells with bilymphoid capacity,[33] no such common lymphoid progenitors (CLPs) with both B and T cell potential have been identified in embryonic lymphoid tissues in any species. Subsequently, the embryonic lymphoid anlage, thymus, bone marrow (in mammals) and the bursa of Fabricius (in avian) become colonized by the progenitor cells. The exact nature, and especially in the case of T cell precursors, the degree of lineage commitment of the seeding cells, are not fully understood.

Among the true and irreversible signs for lymphoid fate adoption are the expression of recombination activating genes, *RAG-1* and *RAG-2*, and the resulting rearrangements of the antigen receptor gene loci. The developing B and T cell precursors also undergo a characteristic sequence of phenotypical and morphological changes. The productively rearranged (V_H to D_H-J_H) immunoglobulin heavy chain genes (*IgH*) make up, together with the surrogate light chain gene products, lambda5 and VpreB, as well as with the associated signal transducers mb-1(Igα) and B29 (Igβ), the preBCR complex. Following a phase of rapid proliferative expansion, the preB cells rearrange either their kappa or lambda light chain genes (*IgL*) to form immature surface IgM positive B cells. In an analogous manner, the early murine double negative (CD4⁻CD8⁻) T cell precursors lacking also the CD3/TCR complex express the RAG proteins to rearrange their *TCRβ*-chain gene loci. Productively made TCRβ is then expressed together with the surrogate α-chain, preTα, (equivalent to the surrogate light chain in preB cells) to form the preTCR. This process called β-selection allows the preT cells to undergo proliferation, to rearrange the *TCRα*-chain genes and finally to become CD8⁺CD4⁺ double positive precursors. Finally, to generate tolerance the antigen receptors of differentiating lymphocytes are subjected to selective screenings to eradicate potentially harmful self-reactive precursors.

In birds, the lymphopoietic system has several important unique features and differences, some of which allow important experimental advantages compared to mammals. As the avian embryos are easily accessed, the first classical experiments to demonstrate the intraembryonic origin for vertebrate lymphoid progeny were performed using grafted chick-quail as well as chick-chick yolk sac embryo chimeras (see chapter by Jaffredo).[16,17] Additionally, the avian lymphopoietic system has a dichotomic nature, as development of the B cell compartment and the Ig repertoire expansion are, separately from T cells, promoted in a specialized organ, called the bursa of Fabricius.[30] Bursa contains follicles that become colonized by intraembryonic mesenchyme-derived and already B lineage-committed progenitors harboring rearranged *Ig* genes.[27,31] Colonization occurs during a single developmental window between embryonic days 8 and 14.[34] In contrast to mammals, a single variable (*V*) heavy (*H*) and light (*L*) chain gene is initially rearranged with a single J element (DJ_H and J_L) to produce the prebursal primary B cell pool. Following bursal homing, these progenitors exploit a unique process called gene conversion to diversify the originally simple Ig repertoire.[32,35] Finally, as chicks, surgically bursectomized early in ontogeny, are able to develop B cells with narrowed Ig repertoire, it has been concluded that B cell development starts independently of bursa.[36] Therefore, the primary role of the bursa of Fabricius is to facilitate Ig repertoire generation through gene conversion and allow proliferative expansion.

In sharp contrast to the avian B cell progenitors with "prehoming" antigen receptor rearrangements, the avian T cell precursors retain their TCR genes in germ line configuration upon homing to thymus.[37] Still, these prethymic cells originating from the intraembryonic para-aortic mesenchyme have been shown to have T cell progenitor characteristics.[29] In contrast to B cell development, the avian T lymphopoiesis is more reminiscent to that of mammals. Nevertheless, a few differences and characteristic features exist in the avian. First, $\gamma\delta$ T cells represent a large fraction of peripheral blood lymphocytes in the avian. In addition, the avian thymus receives three temporally distinct colonizing waves of progenitors. Also, the putative avian homologs of the surrogate Ig light chain genes and the $preT\alpha$, and hence the corresponding preBCR/TCR, have not been documented.

Molecular Control of T Lymphopoiesis

Generation of null mutant embryos has critical limitations considering the interpretations of the resulting phenotypes. The detected developmental arrests provide only evidence of the first step, where a given transcription factor is required. Also, embryonic lethality occasionally brings further complexity to the phenotypic analyses. Although these difficulties more often involve the early acting factors like SCL and Lmo2, they also affect other factors, later proved to play more restricted functions. For example, the third member of the GATA family, GATA-3, is widely expressed during embryogenesis and the corresponding null mutation causes premature death due to severe embryonic hemorrhages.[38] Together with Lmo2 also GATA-3 has been shown to associate with early hematopoietic foci.[39] Its role, however, specifically in T cell development, was discovered in an attempt to circumvent the death-causing abnormalities in null embryos. *GATA-3*-deficient embryonic stem cells were injected into *RAG-2* null blastocysts to generate chimeras, where the contribution of the injected null cells to lymphoid compartments was analysed in a genetically alymphoid background.[40,41] It became apparent that the *GATA-3*-targeted ES cells were capable of giving rise to all the other cell types except cells of the T cell system (Table 1, Fig. 1).

In addition to lethal phenotypes, functional redundancy and overlapping expression profiles have made it difficult to assess the role of individual regulatory factors in lymphocyte development. In line with GATA-3, also Tcf-1 and Lef-1, both members of the high mobility group (HMG) family, are widely expressed in early ontogeny, but have still been shown to display critical lymphoid-restricted functions.[42] *Tcf-1*-targeted mice develop incomplete block in thymocyte development with reduced numbers of immature single positive (ISP) thymocytes (Table 1, Fig. 1). *Lef-1*-deficiency causes aberrant tooth, hair and central nervous system development. Although both factors are known to bind the $TCR\alpha$ enhancer, expression of this cell surface receptor component is unaffected both in *Tcf-1* and *Lef-1* null thymocytes. This has led to an idea of compensatory functions, and therefore double mutants have been generated to address the question, whether Tcf-1 and Lef-1 could exert true functional redundancy. Based on the results obtained, development of the double deficient thymocytes was completely arrested (from ISP to DP).[43] In support of these overlapping roles, it has earlier been demonstrated that the avian Tcf ortholog represents a common ancestor for these two related factors.[44] The exact targets and molecular mechanism for their T lymphoid roles have nevertheless remained unclear, as also the TCR signaling routes are found intact in the double mutant thymocytes.[43] Subsequent analyses of the *Lef-1/Tcf-1*-targeted mutants, however, have revealed similar paraxial mesodermal defects as in *Wnt3a* mutant mice. Indeed, several lines of evidence across species have to date documented the central role of Tcf/Lef factors as nuclear targets of Wnt signaling pathways.[42] These findings extending from *Drosophila melanogaster* to man have opened a new era for understanding, how cytoplasmic signaling is integrated and mediated to nuclear effectors that control early morphogenesis, embryonic axis formation, cell fate decisions and also tumorigenesis.

Molecular Control of B Lymphopoiesis

In analogy with the synergistic role of the Tcf-1/Lef-1 proteins in T cell development, also B lymphopoiesis relies on a network of gene regulators that act in concert to promote B cell fate adoption and lineage progression.[22,45,46] E2A, a member of the bHLH transcription factors, generates two widely expressed splice forms E12 and E47 that both play essential and synergistic roles in B cell development.[22] Knock-out mice display blocked B cell development at a stage prior to *Ig* gene rearrangements (Table 1, Fig. 1). However, when selectively disrupted, *E12*-deficient mice seem to have a slightly later developmental defect in comparison to the B cell defects found in *E47* null littermates. Interestingly, as levels of PU.1 are critical for the decision between macrophage vs B cell fate,[22] forced E12 in turn has been shown to make a macrophage cell line adopt B lineage characteristics. From these studies it has been found that E2A drives the expression of another central regulator, EBF (Fig. 1) (Early B cell Factor).[45] These factors are then required for the expression of multiple B lineage-affiliated genes like *lambda5, VpreB, mb1, B29* and the *RAG* genes.[45] EBF-disrupted phenotype strongly resembles that of *E2A*-deficient mice, and experiments with *E2A/EBF* heterozygous double mutants have revealed their strong collaborative action in promoting the early steps leading to B lineage restriction.

Recent results have documented that neither EBF nor E2A is sufficient to B cell commitment. Indeed, Pax5 (or BSAP), a downstream target of EBF/E2A, has been shown to function as a true B cell identity or master commitment factor.[46] *Pax5*-deficiency arrests B cell development at an early pro-B cell stage and later than *EBF* or *E2A* disruptions (Table 1, Fig. 1). Strikingly, IL-7 depletion induces these *Pax5* mutant pro-B cells to adopt clear characteristics of a multipotential hematopoietic precursor cell.[47] With appropriate cytokine mixtures these cells start to differentiate along osteoclast, dendritic cell, NK cell and granulocyte pathways and have been able to reconstitute the T cell system in *RAG-2*-deficient hosts.[47] In gain-of-function experiments, it has been shown that, when Pax5 expression is rescued via retroviral transduction, normal B cell differentiation can be restored.[47] Conclusion of these results depicts that the role of Pax5 is to irreversibly restrict precursors to the B lineage at the expense of alternate fate choices. In support of this, both Pax5 and EBF have been documented to be active already at the earliest steps of avian B lymphopoiesis.[30] This early expression and ability to shut down other developmental options fits to the multilineage priming concept,[48] but at the same time, it has raised the question of the molecular function of Pax5-mediated repression. One solution to this question was obtained, when Pax5 was demonstrated to interact with Grg4, a member of the Groucho family of transcriptional corepressors.[46] It is currently suggested that Pax5 recruits Groucho and perhaps other related inhibitory factors to repress genes that favor alternate lineage programs. In the case of B cell lineage fate control, the exact molecular mechanism is still not fully understood. It is nevertheless of importance to notice that in other systems Groucho actions have been linked to histone deacetylases with chromatin condensing and gene silencing.[46] Whether chromatin remodeling is required to render the other genetic programs, except the Pax5-favored B lymphoid fate, into a transcriptionally silent state, remains to be resolved. Interestingly, however, conditional gene targeting as well as gain-of-function studies have recently suggested that Notch1, an evolutionarily conserved cell fate regulator, promotes T cell development at the expense of B cell fate.[46] According to this model, Notch1 that is a membrane-bound factor receives its signal exclusively in the thymic environment and prevents E2A activity in the thymus. Therefore, thymic Notch signals instruct T cell fate as a secondary choice, whereas the bone marrow niche lacking Notch signaling allows B cell development as a primary and default outcome. Whether this Notch1-mediated fate decision influences the elusive bipotential lymphoid progenitors or earlier hematopoietic precursors is not known.

The Ikaros Family of Transcription Factors

Ikaros represents the founding member of a zinc finger transcription factor family with central functions in the development and function of the hematopoietic system.[26,49] Identification of the *Ikaros* gene resulted from attempts to define critical molecules involved in T cell lineage specification.[50] As Ikaros was shown be able to drive in vitro reporter gene activity from the CD3δ enhancer, it was originally suggested to function as a T cell commitment factor. Yet, further studies, as described below, have shown it to exert much more diverse effects in the hematopoietic system. *Ikaros* comprises seven exons encoding two separate carboxy- and amino-terminal regions that together contain six Krüppel-like cysteine/histidine-based zinc finger domains (Fig. 2).[49] The four N-terminal motifs mediate high-affinity DNA recognition and binding, whereas the more distal domains are required for dimer formation within the Ikaros family. The primary Ikaros transcript undergoes alternate mRNA processing to produce a functionally diverse set of protein isoforms that differ in their N-terminal zinc finger composition and thus in their DNA binding properties.[49] Structurally Ikaros is most homologous to and has similarities with the *Drosophila* Hunchback segmentation gap protein. Subsequent studies have to date revealed at least three additional members of the *Ikaros* family: *Aiolos*, *Helios* and *Dedalos* and possibly also *Pegasos*.[51-53] These highly related genes have most likely arisen via gene duplication events of a common ancestor to generate a functionally diverse family of gene regulators. Most recent findings have placed members of this family into critical positions in regulating gene activity during hematopoietic lineage specification, differentiation and proliferation. More importantly, Ikaros family has proven out to be among the first identified groups of gene regulators coupling the general chromatin metabolism and gene accessibility control with the lineage-affiliated gene expression programming in lymphocytes.

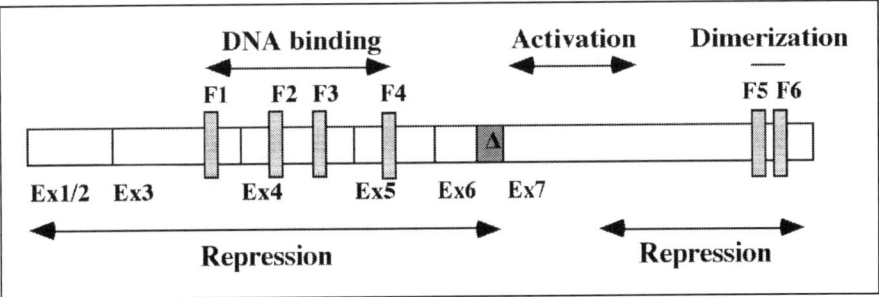

Figure 2. Functional domains and interaction sites of Ikaros. Zinc finger domains F1 to F4 are required for high affinity DNA recognition and binding. C-terminal zinc fingers F5 and F6 are critical for hetero- and homodimerization within the Ikaros family. Regions involved in activatory and repressive properties are indicated by horizontal arrows. The distal segment of exon six is shown in dark color to indicate a 10-amino-acid deletion (Δ) frequently observed in leukemic lymphoblasts.

Contribution of Ikaros to the Hematopoietic Compartments

To elucidate the contribution of Ikaros proteins to the development of the hematopoietic and lymphopoietic systems, two different strategies to inactivate *Ikaros* in mouse germ line have been exploited (Table 1).[49] Targeting of the C-terminal zinc finger domains in the last exon generate unstable and dimerization-deficient Ikaros proteins and hence result in an *Ikaros* null outcome. These animals are devoid of fetal T and B cells as well as adult B cells. After birth however, a small number of T cell precursors can be found in the thymus. Interestingly, these postnatal precursors are predominantly adopting the CD4 T helper phenotype and cannot generate progeny for NK or thymic dendritic cells (DC). For an unknown reason γδ T cells are selectively affected, as mucosal and thymic γδ T cells are present, but the fetal epidermal γδ T cells as well as the gut-associated γδ T cells are absent. The myeloerythroid branch in turn is developing rather normally in *Ikaros* null animals, apart from a block in the terminal maturation of granulocytes. Another *Ikaros* mutation deleting the N-terminal zinc finger domains leads to production of Ikaros proteins that are incapable of DNA binding but still retain intact dimerization and protein interaction motifs. Through dominant negative (DN) influence these mutant proteins can sequester other interacting proteins into transcriptionally inactive complexes unable to bind DNA. In comparison to the *Ikaros* null animals, homozygous DN mutants display an even more profound defect and lack all fetal and adult lymphoid lineages and fail to develop thymic DC and mature granulocytes. Their bone marrow is highly hypocellular and the thymus, although histologically identifiable, is markedly reduced in size and totally devoid of T cells. Of the secondary lymphoid structures, all the peripheral lymph nodes and gut-associated lymphoid patches are absent in the mutants. In sharp contrast to the lack of lymph nodes and the severely reduced marrow hematopoiesis, animals develop severe splenomegaly as a result of markedly enhanced extramedullary myeloerythropoiesis. Given the central immunocompromising defects affecting both the innate and adaptive immune surveillance, DN mutant animals are highly prone to polymicrobial infections and subsequently die during their first month.

Ikaros and Hematopoietic Stem Cells

It originally appeared that murine Ikaros is first expressed in the E9.5 fetal liver.[50] Later, it became apparent that Ikaros can be detected already from the E8 extraembryonic yolk sac.[49] Therefore, the effects of *Ikaros* mutations on the hematopoietic precursor and stem cell compartments were analysed in more detail using both in vitro and in vivo colony-forming assays.[54] *Ikaros* null animals had a dramatically lower numbers of long-term repopulating HSCs when compared to the wild-type littermates. The DN mutants were totally devoid of this repopulation activity. In line with the HSC defects, the more differentiated CFU-S (colony-forming unit-spleen) and the BFU-E (burst-forming unit-erythroid) were also affected. Later erythromyeloid precursors, however, were more intact in the targeted animals. Both *Ikaros* mutations impaired the expression of c-kit and flk2. Yet, the cellular consequences caused by the reduced levels of these two cell surface receptors are still unresolved. In addition to the defects in adult HSC compartments, also embryonic precursors from E10 AGM and yolk sac were similarly affected when the CFU-S contents were assayed.

Ikaros and Thymocyte Development

The generated *Ikaros* mutations cause only minor changes in the levels of SCL, PU.1 and GATA-2 transcription factors.[54] This is most likely a secondary effect and is due to the relative alterations in the precursor population sizes. In contrast, however, the T cell lineage-specifying factor GATA-3 is totally absent in the DN mutants.[54] In a similar way, lack of GATA-3 expression may be due to the missing T lymphopoiesis, or alternatively, it may be a direct consequence of the DN *Ikaros* mutation.[49] During thymocyte development *Ikaros* deficiency dissociates differentiation from the normally occurring phases of proliferative expansions that are mediated by the preTCR and TCR receptor complexes.[55] First, evidence from studies using *Ikaros* null X *RAG-1* double mutants showed that absence of Ikaros can promote differentiation in RAG-1 negative background, where thymopoiesis is normally blocked at the double negative stage due to the absence of a functional preTCR. Moreover, the observed transition into double positive stage occurred without the normally required proliferation. Second, *Ikaros* null X *TCRα* double mutants can differentiate from double positive to immature single positive thymocytes, a process normally blocked in *TCRα* mutants. Thus, positive selection seems to occur without TCR engagement in a similar way as the earlier β-selection occurred without any preTCR signaling. Taken together, Ikaros seems to set at least two distinct bottlenecks for critical steps in thymocyte development. In normal thymocytes, Ikaros could thus function as a gate-keeper for preTCR and TCR-delivered signals required for further development.

Ikaros: A Threshold for T Cell Function

Subsequent analyses of animals heterozygous for the dominant negative *Ikaros* mutation revealed that they undergo augmented T lymphoproliferation and finally succumb to lymphomas and leukemias.[49] The observed clonal outgrowth is accompanied by loss of the remaining wild-type *Ikaros* allele and can be first detected within the thymic T cell populations. Transitional state intermediates between the double and single positive thymocyte subsets start to predominate. As described above, this accumulation of expanding thymocytes is likely to be due to the significantly decreased Ikaros-based proliferation thresholds. Thymocytes have intact TCRs that are, without Ikaros interference, able to mediate proliferative expansion resulting in neoplastic transformation. In support of this, certain peripheral T cell subsets, e.g., splenocytes, are autoproliferative as they can expand without TCR engagement.[49] By the age of 3 months, mutant animals develop enlarged lymph nodes and severe splenomegaly as well as have lymphoblastic infiltrates in multiple organs including bone marrow. Circulating leukemic lymphoblasts and the tumor-forming cells are of clonal origin, as they display the same TCR specificity. In conclusion, dominant negative *Ikaros* mutation generates variants that bind normal Ikaros proteins and other interacting factors into inert complexes. The resulting decrease in the overall Ikaros level lowers the Ikaros-mediated preTCR/TCR-signaling thresholds thereby allowing illegitimate thymocyte proliferation and leukemogenesis. Nevertheless, a functional TCR is required for leukemogenesis, as no DN mutation-induced malignancies develop in *RAG*-deficient background.[55] Therefore, *Ikaros*-deficiency is sufficient to promote thymocyte differentiation without TCR, but is unable to drive malignant transformation in the absence of TCR. In addition to the heterozygous DN mutation, also *Ikaros* null animals develop clonal expansion and malignant out-growth.[49]

Aiolos: Setting Brakes for B Cells

Since the DN *Ikaros* mutation causes via dominant negative interference a more severe outcome compared to the null mutation, interacting Ikaros partners are likely to exist. As a result, Aiolos, the second Ikaros family member, has been isolated and shown to share similar structural and functional properties.[51] By contrast to Ikaros, Aiolos is not expressed in the early hematopoietic precursors or stem cells, but is actively present in the later stages of lymphocyte development (Table 1, Fig. 1).[51,52] Although murine Aiolos readily dimerizes with the Ikaros variants, Aiolos itself has not been demonstrated to generate any splice variants and is therefore suggested to form only one type of Aiolos-Aiolos homodimers.[51] Contribution of Aiolos to the lymphopoietic system was analysed by targeting a portion of the last *Aiolos* exon.[56] Strikingly, the resulting null mutants revealed that Aiolos performs highly analogous functions in the B cell development, as Ikaros has been shown to carry out in the T lymphopoiesis (Table 1). Bone marrow preB cells are increased, but B cells from the peritoneum and marginal zone as well as the recirculating marrow-derived populations are reduced. There was also a marked increase in the number of activated peripheral B cells and appearance of splenic germinal centers without immunization. B cell tolerance was lost, as immune responses were elicited and autoantibodies produced in the absence of foreign antigens. Therefore, *Aiolos*-deficiency seems to lower the BCR-signaling threshold and disrupt the normal selective processes. Importantly, *Ikaros*-deficiency facilitates in a similar way the progression of double positive thymocytes into single positive state without the normally required TCR-mediated expansion.[55] As a result, T cells with autoreactive TCR can develop. In line with the *Ikaros*-deficient T cells, also *Aiolos* null B cells hyperproliferate upon in vitro stimulation. Finally, *Aiolos* mutant animals develop via the aberrant B lymphoproliferation clonal B cell lymphomas. *Ikaros-Aiolos* double mutants display even more enhanced T and B cell expansions, suggesting a synergistic role for these proteins in setting signaling thresholds.[26] Thus, both Aiolos and Ikaros function as gate-keepers in antigen receptor-mediated signaling involving differentiation, proliferation, selection processes and terminal maturation as well as tumor suppression.

Ikaros and Aiolos Running the Lymphoid Chromatin Modifying Machinery

One of the first insights into the lymphoid-related repressor mechanisms of Ikaros came from studies showing that Ikaros makes functionally relevant associations with transcriptionally inert centromeric heterochromatin.[57] Using an immuno-FISH (fluorescence in situ hybridization)-based approach it was demonstrated that within the nuclei of B lymphocytes Ikaros colocalizes with T cell-specific genes e.g., *CD4* and *CD8* to centromeric regions. Moreover, during B cell development, genes not any more required for the terminal maturation, e.g., *lambda5*, are deposited with Ikaros into silent chromatin loci. In addition, the number of centromeric Ikaros foci seems to increase during the synthesis phase of cell cycle, which would indicate along with other studies a role for Ikaros in DNA replication control. In conclusion, these studies suggest a scenario, where genes could be in a dynamic flux between transcriptionally permissive and inhibitory nuclear contexts depending on how these target genes are required. Accordingly, Ikaros would function as a molecular bridge that carries genes between these two states, and not only during one individual cell cycle, but also during the development and function of B lymphoid cells.

In the nuclei of lymphocytes, Ikaros and Aiolos function as lineage- and cell-specific recruiters of the various chromatin remodeling complexes (Fig. 3).[26,58] Indeed, yeast two-hybrid screens with Aiolos baits originally led to the identification and cloning of the murine *Mi-2* chromatin remodeling ATPase homologs.[58] Hence, the repressor mechanisms by which Ikaros family factors target inert genes to heterochromatic sites involve chromatin modifications. Upon lymphocyte activation, Ikaros and Aiolos direct the formation of large 2 MD multisubunit complexes that contain components of the NuRD (i.e., Mi-2 and HDACs) and other interacting factors.[58] In cycling cells, these structures enter the previously characterized toroidal structures around the centromeric heterochromatin. In addition to NuRD, a smaller fraction of Ikaros is also capable of making interactions and similarly sized assemblies with Brg-1, a member of the SWI/SNF chromatin remodeling complexes. By contrast to Ikaros-NuRDs, these complexes remain diffusively scattered in the nucleus throughout the T cell cycle, suggesting that they may reside in the euchromatic regions. In line with this hypothesis, it has previously been shown that members of the SWI/SNF remodeling factors might, by contrast to NuRDs, be involved in rendering chromatin into an accessible state.[59]

In further yeast two-hybrid studies with Ikaros as a bait, it has been shown that Ikaros and also Aiolos interact with Sin3 repressor complexes.[60] Ikaros has several interaction regions for

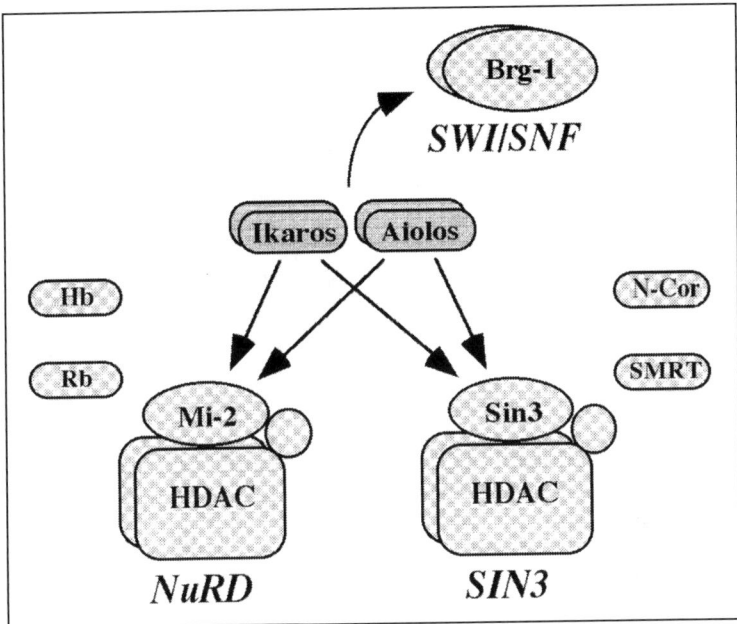

Figure 3. Transcription regulators interacting with the chromatin remodeling complexes. NuRD (nucleosome remodeling histone deacetylase complex) and Sin3 deacetylate histones and render target gene loci into condensed and transcriptionally silent state. SMRT and N-Cor are corepressor molecules involved in Sin3-mediated transcriptional repression. In addition to Ikaros and Aiolos, also the *Drosophila* Hunchback (Hb) is known to interact with dMi-2. Similarly, the Retinoblastoma (Rb) tumor suppressor is able to make interactions with the NuRD complex and function as a corepressor. Ikaros has also been shown to associate with Brg-1 (Brahma group) a member of the SWI/SNF complex that is suggested to remodel target gene loci into an accessible and transcriptionally active state.

Sin3 that are independent of the DNA recognition and dimerization motifs. In contrast to the earlier reporter assays, Ikaros and Aiolos were now found to act as strong transcriptional repressors when carried to DNA by a heterologous DNA binding domain of Gal4. This repression is not dependent on Ikaros-mediated DNA binding. Furthermore, Ikaros isoform 6 lacking all the DNA recognizing motifs functions as the strongest repressor. The detected repression depends on Sin3 and has promoter- and cell type-specific features, as distinct promoters and cell lines displayed different amounts of promoter activity repression. Although clearly involved in gene silencing, Ikaros-Sin3 complexes fail to enter, like Ikaros-NuRD, into the heterochromatic sites during T cell activation. Moreover, these studies conclude that the detected repression does not require direct Ikaros- or Aiolos-mediated DNA binding, as the chromatin modifying complexes were carried to DNA by a heterologous DNA binding motif of Gal4. In other studies, however, using ectopic Ikaros expression in 3T3 fibroblasts, it has been demonstrated that direct DNA binding of Ikaros is crucial for the protein complexes to enter the pericentromeric heterochromatin.[61] Two central zinc fingers in the N-terminal DNA binding region were shown to be critical for recognition of Ikaros binding sites present in the centromeric regions. Interestingly, the Mi-2 component of the NuRD complex was not significantly associated with Ikaros in 3T3 fibroblasts and was not required for centromeric Ikaros targeting. There are also studies showing Mi-2/HDAC-independent repressor mechanisms of Ikaros.

Collectively, several models of Ikaros/Aiolos-mediated gene regulation and chromatin remodeling activities can be drawn and hypothesized according to the results obtained to date.[26] First, it could be suggested that members of the Ikaros family recruit distinct chromatin remodeling machineries, i.e., NuRD, Sin3 and SWI/SNF, into target gene loci in a developmental stage-, cell cycle- and cell-specific manner. NuRD and Sin3 complexes with HDAC activity would drive chromatin condensation and deacetylation and thus render associated gene loci into silence. Ikaros/Aiolos-NuRD has been suggested to function in maintaining the repressed state through cell divisions, whereas Ikaros/Aiolos-Sin3 may play a role in silencing active genes by chromatin condensation. Ikaros would therefore allow stable transfer of genetic programs via heritable gene control into the cell progeny emerging as a result of cell divisions. A hypothetical role for the SWI/SNF-containing Ikaros/Aiolos complexes in turn might be to activate gene expression by remodeling chromatin into a more accessible configuration. Alternatively, studies revealing direct binding of Ikaros to the centromeric foci could indicate that Ikaros first binds DNA and then recruits components of the chromatin modifying machinery to regulate gene activity. Another model has suggested that centromeric sites store Ikaros proteins that subsequently could be used elsewhere in the nucleus for chromatin remodeling purposes.[61]

The Ikaros Family in Avian Lymphohematopoiesis

It was first shown in the avian that Ikaros is expressed before the colonization of the primary lymphoid organs (Fig. 4).[62,63] Thereafter Ikaros transcripts are detected in prethymic

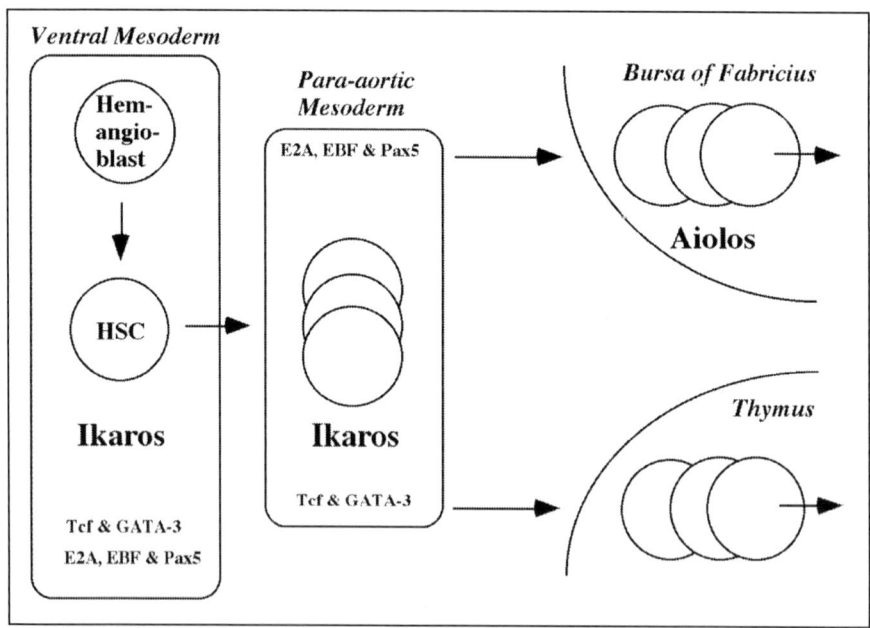

Figure 4. Transcription factors in the early avian hematopoietic development. Ikaros expression starts from the second day of incubation onwards, whereas Aiolos is first detected in the avian bursa of Fabricius. At E7 Ikaros is expressed in para-aortic cells having T cell progenitor capacity. Tcf and GATA-3 transcription factors are expressed during early ontogeny and continue to be active during thymocyte development. In a similar way, E2A, EBF and Pax5 become detectable in early embryonic tissues and are expressed together with Aiolos in embryonic bursal cells. HSC: hematopoietic stem cell.

para-aortic progenitors that have been shown to have capacity to give rise to T cell progeny.[29] In line with Ikaros, factors involved in B lymphopoiesis, i.e., E2A, EBF and Pax5, as well as key regulators of T cell development, i.e., GATA-3 and Tcf, are expressed early in ontogeny (Fig. 4).[30,37] By contrast, the avian Aiolos ortholog is first detected in the embryonic bursa of Fabricius.[64] Importantly, Aiolos expression increases concomitantly with the number of sIgM positive cells. Both Ikaros and Aiolos are evolutionarily highly conserved across avian and mammalian species. Studies in the avian have provided the first evidence that also Aiolos generates alternatively spliced isoforms.[63,64] Importantly, Aiolos variants are not only expressed in avian lymphocytes, but also in normal and leukemic human B lymphocytes.[65] In addition, two of the new variants lack functionally central DNA binding domains and might therefore have a dominant negative influence on other Ikaros family members. Indeed, structurally similar Ikaros isoforms with dominant negative effects have been demonstrated to associate to human acute lymphoblastic leukemias. In summary, identification of novel Aiolos variants renders the chromatin remodeling machinery directing lymphoid gene expression more complex by bringing new components to the regulatory assemblies. Taken together, characterization of transcription factors as well as their new isoforms offers new insights into the understanding of stem cell biology and lymphohematopoiesis and may thereby also give molecular cues to the pathogenesis of both immunological and hematological diseases.

References

1. Choi K, Kennedy M, Kazarov A et al. A common precursor for hematopoietic and endothelial cells. Development 1998; 125:725-732.
2. Jaffredo T, Gautier R, Brajeul V et al. Tracing the progeny of the aortic hemangioblast in the avian embryo. Dev Biol 2000; 224:204-214.
3. Jaffredo T, Gautier R, Eichmann A et al. Intraaortic hemopoietic cells are derived from endothelial cells during ontogeny. Development 1998; 125:4575-4583.
4. Pardanaud L, Dieterlen-Lièvre F. Manipulation of the angiopoietic/hemangiopoietic commitment in the avian embryo. Development 1999; 126:617-627.
5. Pardanaud L, Luton D, Prigent M et al. Two distinct endothelial lineages in ontogeny, one of them related to hemopoiesis. Development 1996; 122:1363-1371.
6. Caprioli A, Jaffredo T, Gautier R et al. Blood-borne seeding by hematopoietic and endothelial precursors from the allantois. Proc Natl Acad Sci USA 1998; 95:1641-1646.
7. Eichmann A, Corbel C, Pardanaud L et al. Hemangioblastic precursors in the avian embryo. Curr Top Microbiol Immunol 2000; 251:83-90.
8. Lampisuo M, Karvinen J, Arstila TP et al. Intraembryonic haemopoietic cells and early T cell development. Scand J Immunol 1995; 41:65-69.
9. Wood HB, May G, Healy L et al. CD34 expression patterns during early mouse development are related to modes of blood vessel formation and reveal additional sites of hematopoiesis. Blood 1997; 90:2300-2311.
10. Eichmann A, Corbel C, Nataf V et al. Ligand-dependent development of the endothelial and hemopoietic lineages from embryonic mesodermal cells expressing vascular endothelial growth factor receptor 2. Proc Natl Acad Sci USA 1997; 94:5141-5146.
11. Takakura N, Watanabe T, Suenobu S et al. A role for hematopoietic stem cells in promoting angiogenesis. Cell 2000; 102:199-209.
12. Porcher C, Swat W, Rockwell K et al. The T cell leukemia oncoprotein SCL/tal-1 is essential for development of all hematopoietic lineages. Cell 1996; 86:47-57.
13. Gering M, Rodaway ARF, Göttgens B et al. The SCL gene specifies haemangioblast development from early mesoderm. EMBO J 1998; 17:4029-4045.
14. Olson EN. MyoD family: A paradigm for development? Genes Dev 1990; 4:1454-1461.
15. North T, Gu TL, Stacy T et al. Cbfa2 is required for the formation of intra-aortic hematopoietic clusters. Development 1999; 126:2563-2575.
16. Dieterlen-Lièvre F. On the origin of haematopoietic stem cells in the avian embryo: An experimental approach. J Embryol Exp Morphol 1975; 33:607-619.

17. Lassila O, Eskola J, Toivanen P et al. The origin of lymphoid stem cells studied in chick yolk sac-embryo chimaeras. Nature 1978; 272:353-354.

18. Cumano A, Dieterlen-Lièvre F, Godin I. Lymphoid potential, probed before circulation in mouse, is restricted to caudal intraembryonic splanchnopleura. Cell 1996; 86:907-916.

19. Medvinsky A, Dzierzak E. Definitive hematopoiesis is autonomously initiated by the AGM region. Cell 1996; 86:897-906.

20. Lassila O, Martin C, Toivanen P et al. Erythropoiesis and lymphopoiesis in the chick yolk sac-embryo chimeras: Contribution of yolk sac and intraembryonic stem cells. Blood 1982; 59:377-381.

21. Sieweke MH, Graf T. A transcription factor party during blood cell differentiation. Curr Opin Gen Dev 1998; 8:545-551.

22. Glimcher LH, Singh H. Transcription factors in lymphocyte development - T and B cells get together. Cell 1999; 96:13-23.

23. Ohmura K, Kawamoto H, Fujimoto S et al. Emergence of T, B, and myeloid lineage-committed progenitors in the aorta-gonad-mesonephros region of day 10 fetuses of the mouse. J Immunol 1999; 163:4788-4795.

24. Tsang AP, Fujiwara Y, Hom DB et al. Failure of megakaryopoiesis and arrested erythropoiesis in mice lacking the GATA-1 transcriptional cofactor FOG. Genes Dev 1998; 12:1176-1188.

25. Nerlov C, Querfurth E, Kulessa H et al. GATA-1 interacts with the myeloid PU.1 transcription factor and represses PU.1-dependent transcription. Blood 2000; 95:2543-2551.

26. Cortes M, Wong E, Koipally J et al. Control of lymphocyte development by the Ikaros gene family. Curr Opin Immunol 1999; 11:167-171.

27. Lassila O, Eskola J, Toivanen P. Prebursal stem cells in the intraembryonic mesenchyme of the chick embryo at 7 days of incubation. J Immunol 1979; 123:2091-2094.

28. Delassus S, Titley I, Enver T. Functional and molecular analysis of hematopoietic progenitors derived from the aorta-gonad-mesonephros region of the mouse embryo. Blood 1999; 94:1495-1503.

29. Lampisuo M, Liippo J, Vainio O et al. Characterization of prethymic progenitors within the chicken embryo. Int Immunol 1999; 11:63-69.

30. Nieminen P, Liippo J, Lassila O. Bursa of Fabricius. Encyclopedia of Life Sciences / ©. Macmillan Publishers Ltd, Nature Publishing Group, 2001.

31. Mansikka A, Sandberg M, Lassila O et al. Rearrangement of immunoglobulin light chain genes in the chicken occurs prior to colonization of the embryonic bursa of Fabricius. Proc Natl Acad Sci USA 1990; 87:9416-9420.

32. Weill JC, Reynaud CA. The chicken B cell compartment. Science 1987; 238:1094-1098.

33. Kondo M, Weissman IL, Akashi K. Identification of clonogenic common lymphoid progenitors in mouse bone marrow. Cell 1997; 91:661-672.

34. Le Douarin NM, Houssaint E, Jotereau FV et al. Origin of hematopoietic stem cells in embryonic bursa of Fabricius and bone marrow studied through interspecific chimeras. Proc Natl Acad Sci USA 1975; 72:2701-2705.

35. Lassila O, Alanen A, Lefkovits I et al. Immunoglobulin diversification in embryonic chicken bursae and in individual bursal follicles. Eur J Immunol 1988; 18:943-949.

36. Mansikka A, Jalkanen S, Sandberg M et al. Bursectomy of chicken embryos at 60 hours of incubation leads to an oligoclonal B cell compartment and restricted Ig diversity. J Immunol 1990; 145:3601-3609.

37. Liippo J, Koskela K, Lassila O. Prethymic progenitors from the avian para-aortic mesoderm express GATA-3 and distinct chTcf isoforms but still lack T-cell receptor-γ rearrangements. Scand J Immunol 2000; 52:502-509.

38. Pandolfi PP, Roth ME, Karis A et al. Targeted disruption of the GATA-3 gene causes severe abnormalities in the nervous system and in fetal liver haematopoiesis. Nat Genet 1995; 11:40-44.

39. Manaia A, Lemarchandel V, Klaine M et al. Lmo2 and GATA-3 associated expression in intraembryonic hemogenic sites. Development 2000; 127:643-653.

40. Ting C, Olson MC, Barton KP et al. Transcription factor GATA-3 is required for development of the T-cell lineage. Nature 1996; 384:474-478.

41. Kuo CT, Leiden JM. Transcriptional regulation of T lymphocyte development and function. Annu Rev Immunol 1999; 17:149-187.

42. Clevers H, van de Wetering M. TCF/LEF factor earn their wings. Trends Genet 1997; 12:485-489.
43. Okamura RM, Sigvardsson M, Galceran J et al. Redundant regulation of T cell differentiation and TCRα gene expression by the transcription factors LEF-1 and TCF-1. Immunity 1998; 8:11-20.
44. Castrop J, Hoevenagel R, Young JR et al. A common ancestor of the mammalian transcription factors TCF-1 and TCF-1α/LEF-1 expressed in chicken T cells. Eur J Immunol 1992; 22:1327-1330.
45. Reya T, Grosschedl R. Transcriptional regulation of B-cell differentiation. Curr Opin Immunol 2000; 10:158-165.
46. Busslinger M, Nutt SL, Rolink AG. Lineage commitment in lymphopoiesis. Curr Opin Immunol 2000; 12:151-158.
47. Nutt SL, Heavey B, Rolink AG et al. Commitment to the B lymphoid lineage depends on the transcription factor Pax5. Nature 1999; 401:556-562.
48. Enver T, Greaves M. Loops, lineages, and leukemia. Cell 1998; 94:9-12.
49. Georgopoulos K, Winandy S, Avitahl N. The role of the Ikaros gene in lymphocyte development and homeostasis. Annu Rev Immunol 1997; 15:155-176.
50. Georgopoulos K, Moore DD, Derfler B. Ikaros, an early lymphoid-specific transcription factor and a putative mediator for T cell commitment. Science 1992; 258:808-812.
51. Morgan B, Sun L, Avitahl N et al. Aiolos, a lymphoid restricted transcription factor that interacts with Ikaros to regulate lymphocyte differentiation. EMBO J 1997; 16:2004-2013.
52. Kelley CM, Ikeda T, Koipally J et al. Helios, a novel dimerization partner of Ikaros expressed in the earliest hematopoietic progenitors. Curr Biol 1998; 8:508-515.
53. Hahm K, Cobb BS, McCarty AS et al. Helios, a T cell-restricted Ikaros family member that quantitatively associates with Ikaros at centromeric heterochromatin. Genes Dev 1998; 12:782-796.
54. Nichogiannopoulou A, Trevisan M, Neben S et al. Defects in hemopoietic stem cell activity in Ikaros mutant mice. J Exp Med 1999; 190:1201-1213.
55. Winandy S, Wu L, Wang J-H et al. PreT cell receptor (TCR) and TCR-controlled checkpoints in T cell differentiation are set by Ikaros. J Exp Med 1999; 190:1039-1048.
56. Wang J-H, Avitahl N, Cariappa A et al. Aiolos regulates B cell activation and maturation to effector state. Immunity 1998; 9:543-553.
57. Brown KE, Guest SS, Smale ST et al. Association of transcriptionally silent genes with Ikaros complexes at centromeric heterochromatin. Cell 1997; 91:845-854.
58. Kim J, Sif S, Jones B et al. Ikaros DNA-binding proteins direct formation of chromatin remodeling complexes in lymphocytes. Immunity 1999; 10:345-355.
59. Workman JL, Kingston RE. Alteration of nucleosome structure as a mechanism of transcriptional regulation. Annu Rev Biochem 1998; 67:545-579.
60. Koipally J, Renold A, Kim J et al. Repression by Ikaros and Aiolos is mediated through histone deacetylase complexes. EMBO J 1999; 18:3090-3100.
61. Cobb BS, Morales-Alcelay S, Kleiger G et al. Targeting of Ikaros to pericentromeric heterochromatin by direct DNA-binding. Genes Dev 2000; 14:2146-2160.
62. Liippo J, Lassila O. Avian Ikaros gene is expressed early in embryogenesis. Eur J Immunol 1997; 27:1853-1857.
63. Liippo J, Nera K-P, Kohonen P et al. The Ikaros family and the development of early intraembryonic hematopoietic stem cells. Curr Topics Microbiol Immunol 2000; 251:51-58.
64. Liippo J, Mansikka A, Lassila O. The evolutionarily conserved avian Aiolos gene encodes alternative isoforms. Eur J Immunol 1999; 29:2651-2657.
65. Liippo J, Nera K-P, Lähdesmäki A et al. Both normal and leukemic B lymphocytes express multiple isoforms of the human Aiolos gene. Eur J Immunol 2001; 31:3469-3474.

CHAPTER 5

Yolk Sac Development in Mice

James Palis

Introduction

A functional cardiovascular system is essential for survival and growth of the mammalian embryo, so the hematopoietic, vascular, and cardiac organ systems are the first to emerge in early post-implantation development.[1] During gastrulation, mesoderm cells create the body plan, forming red blood cells in the yolk sac, the heart and aorta in the embryo proper, and a vascular network to connect the two. In this chapter, the development of the hematopoietic system in the yolk sac of the mouse will be reviewed.

Gastrulation and Yolk Sac Formation

Studies in multiple vertebrates, including birds, fish, amphibians, and mammals, indicate that the initial generation of blood cells in the embryo depends on gastrulation and formation of mesoderm cells. In the mouse embryo, implantation occurs at embryonic day 4.5 (E4.5) and gastrulation begins at E6.5, as mesoderm cells traverse the primitive streak to occupy an intermediate position between the primitive ectoderm and visceral endoderm germ layers[2] (Fig. 1A). Cell tracking studies indicate that the mesoderm cells that exit the posterior primitive streak migrate proximally and appose visceral endoderm, extraembryonic ectoderm, and embryonic ectoderm to form the yolk sac, the chorion, and the amnion, respectively (Fig. 1B). Extraembryonic mesoderm cells also form the allantois that contains umbilical vessels connecting the embryo proper to the placenta.[3] All of these extraembryonic structures are essential for the survival of the embryo.[4]

While one third of mesoderm cells crossing the primitive streak migrate to extraembryonic sites, the remaining two thirds migrate to the embryo proper and contribute to formation of paraxial, intermediate, and lateral plate mesoderm. Dorsal lateral plate mesoderm apposed to ectoderm forms somatopleure and gives rise to the lining of the body wall. In contrast, ventral lateral plate mesoderm apposed to endoderm forms splanchnopleure and gives rise to the outer lining of the viscera and contributes to formation of the aorta. Finally, mesoderm cells from the most anterior portion of the primitive streak form definitive endoderm[5] that displaces the visceral endoderm from the distal embryo proper[6] (Fig. 1B). Definitive endoderm cells ultimately contribute to the formation of several intra-embryonic organs, including the gut, lungs, liver, and pancreas.

The yolk sac is a splanchnopleuric organ that maintains a bilayer structure of mesoderm-derived and of visceral endoderm-derived cell layers (Fig. 1B). Survival of the embryo depends on functions provided both by the endoderm and by the mesoderm components of the yolk sac. Visceral endoderm cells of the yolk sac transport and metabolize maternally-derived

Hematopoietic Stem Cell Development, edited by Isabelle Godin and Ana Cumano.
©2006 Eurekah.com and Kluwer Academic / Plenum Publishers.

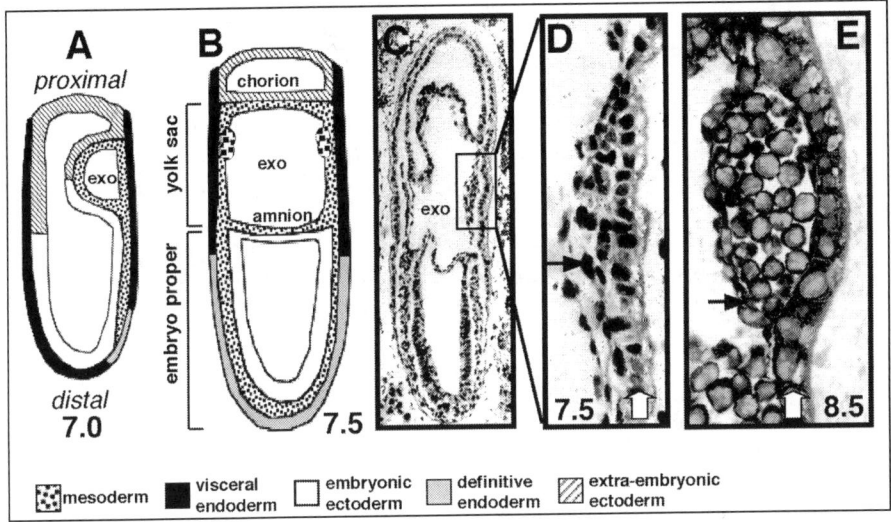

Figure 1. Gastrulation and blood island formation. A) At E7.0, mesoderm cells migrate proximally to line the exocoelomic cavity (exo) and distally to enter the embryo proper. B) By E7.5 (neural plate stage), mesoderm cells associate with visceral endoderm, embryonic ectoderm, and extraembryonic ectoderm to form the yolk sac, amnion, and chorion, respectively. Definitive endoderm cells displace visceral endoderm at the distal embryo proper. C) At E7.5, mesoderm adjacent to visceral endoderm start to form angioblastic cords/mesodermal cell masses (box, D). Higher magnification of angioblastic cord (black arrow) adjacent to cuboidal visceral endoderm cells (white arrow). E) At E8.5, blood islands contain primitive erythroblasts surrounded by endothelial cells (black arrow) adjacent to cuboidal visceral endoderm cells (white arrow).

macromolecules and synthesize serum proteins,[7] functions that are later assumed by the definitive endoderm-derived gut and liver. Mesoderm cells of the yolk sac produce the embryo's first blood cells in blood islands. Pioneering experiments in the chick embryo led to the concept that signals from the visceral endoderm induce the formation of blood and endothelium in adjacent yolk sac mesoderm.[8,9] The importance of visceral endoderm on blood vessel formation in yolk sac mesoderm was also shown in mouse embryos, since blood cells and blood vessels develop in explants of intact yolk sacs, while vessels failed to develop in explants of extraembryonic mesoderm alone.[10] Further support for the role of visceral endoderm in blood cell and blood vessel formation comes from GATA-4-null embryonic stem cell-derived embryoid bodies that lack visceral endoderm and display markedly reduced blood island formation.[11] Tissue recombination experiments in the mouse suggest that these endoderm-inducing signals act during early gastrulation and are diffusible.[12] Subsequent studies indicated that Indian Hedgehog signaling is an important component of this induction.[13,14] While the downstream targets of hedghog signaling are not yet elucidated, BMP4 is a likely candidate. Mouse embryos lacking BMP4 have a paucity of yolk sac blood islands and die in early post-implantation stages because of defects in mesoderm formation.[15] Furthermore, exogenous BMP-4 mediates the development of mesoderm cells and hematopoietic cells in embryonic stem cell-derived embryoid bodies grown under serum free conditions.[16]

Blood Island Formation

Yolk sac blood island development was initially investigated in the early 1900s.[17,18] Phase and electron microscope studies have been performed in several mammalian species, including mouse,[19,20] guinea pig,[21] gerbil,[22] cat,[23] and human.[24] In the mouse, blood islands arise in the yolk sac between E7-7.5, as mesoderm cells adjacent to visceral endoderm proliferate to form angioblastic cords also termed mesodermal cell masses[19] (Fig. 1C, D). Over the next 24 hours, cells of the angioblastic cord differentiate into morphologically distinguishable erythroid cells and these 'blood islands' become surrounded by a layer of endothelial cells (Fig. 1E).

The formation of a primary vascular network from mesoderm cells is termed vasculogenesis.[25] Yolk sac angioblasts coalesce into a primitive vascular network that is ultimately remodeled into a branching system of arteries and veins by E10 in the mouse embryo. Over this same period, angioblasts coalesce in the proximal embryo proper to form the cardiac endothelium, the aorta, and the omphalomesenteric artery that will connect the embryonic and extraembryonic vasculature.[26] Major arteries ultimately become ensheathed with vascular smooth muscle cells that provide strength, stability, and pressure regulation.

Hemangioblast

The intimate temporal and spatial association of blood cells and endothelial cells in yolk sac blood islands led to the hypothesis that blood cells and blood vessels both arise from a common precursor. Murray[27] used the term "hemangioblast" to describe the mesodermal cell masses that give rise to a blood filled vasculature. The concept of a common hemangioblast precursor for both blood and endothelium is supported by the coexpression of a wide range of genes, including Scl, lmo-2, GATA-2, EPO-R, tie, tie2 and PECAM, in both lineages.[28-31] Furthermore, the targeted disruption of several of these genes, including flk-1, Scl, and lmo-2, each results both in hematopoietic and in vascular defects. Flk-1-null embryos fail to form blood islands at E7.5 and die soon afterwards with a complete lack of endothelial cells and blood cells.[32] A subsequent study found that hematopoietic potential was present in E7.5 embryos lacking flk-1, suggesting that flk-1 may function in hemangioblast migration during gastrulation.[33] Mouse embryos lacking Scl and lmo-2, each die between E9.5-10.5 from a complete lacked hematopoietic cells in endothelial-lined yolk sacs.[34-36] However, chimeric mouse studies with Scl-null and lmo-2-null embryonic stem cells each revealed vascular remodeling defects later in gestation suggesting that these genes also function in vascular development.[37,38]

The most direct evidence for the existence of a hemangioblast cell comes from in vitro experiments with cultured embryonic stem (ES) cells. The laboratory of G. Keller identified a unique precursor, termed the Blast-CFC, that gives rise both to hematopoietic progenitors and endothelial cells upon secondary replating.[39] Blast-CFC express the tyrosine kinase receptor flk-1 and are dependent upon its ligand, vascular endothelial growth factor (VEGF), for growth in vitro.[39,40] The transcription factor Scl plays an essential role in the formation of in vitro colonies from Blast-CFC.[41] Flk-1-positive cells isolated from ES cell cultures can give rise to flat sheets of VE-cadherin-positive endothelial cells and round blood cells.[42] It was subsequently determined that single flk-1-positive cells purified from ES cell cultures can give rise to endothelial cells and to vascular smooth muscle cells,[43] suggesting a common origin for both major elements of the vessel wall. Blast-CFC have recently been described during gastrulation in the mouse embryo.[44] These hemangioblast precursors are found predominantly in the posterior primitive streak and can give rise to hematopoietic, endothelial, and smooth muscle cells, but not to other mesodermal derivatives, such as cardiac or skeletal muscle cells.[44] These studies support the concept that the first hematopoietic cells in the mammalian embryo are derived from hemangioblast precursors.

Further support for the close association of the vascular and hematopoietic lineages comes from morphologic studies and transplantation assays suggesting that the hematopoietic stem cells that contribute to long-term hematopoiesis in the adult mouse arise from the aorta beginning at E10.5, when cell clusters are found associated with its endothelial lining.[45-47] The concept of a "hemogenic endothelium" is strengthened by studies in Runx-1-null embryos that lack these cell clusters as well as hematopoiesis in the fetal liver.[48-50]

Primitive Erythropoiesis

The predominant hematopoietic lineage to emerge from yolk sac blood islands is a "primitive" form of erythropoiesis.[17] Nucleated primitive erythroblasts begin to accumulate globin transcripts at E7.5[51,52] and enter the newly formed vasculature of the embryo proper at early somite pair stages with the onset of cardiac function.[53,54] They continue to accumulate hemoglobin for several days reaching a final hemoglobin content of 80 pg/cell.[55-57] Primitive erythroblasts are extremely large cells with cell volumes of 400-800 fl.[58] Primitive red cells display progressive nuclear condensation, ultimately enucleate and circulate for several days as fully mature primitive erythrocytes.[58] The primitive erythroid lineage is unique, because it constitutes the only transient hematopoietic lineage in mammals.

Development of Unipotential Hematopoietic Progenitors

The development of in vitro colony-forming assays in semisolid media led to the identification of unipotential and multipotential hematopoietic progenitors. The earliest recognizable erythroid-specific progenitor is the burst-forming unit erythroid (BFU-E) that in the mouse gives rise to large colonies of red cells after 7-10 days in culture.[59] BFU-E generate colony-forming units erythroid (CFU-E) that rapidly give rise to small colonies of red cells in 2-3 days of in vitro culture.[60] Examination of the early post-implantation mouse embryo for BFU-E and CFU-E led to the discovery of a unique intermediate erythroid progenitor type, the "day 5 CFU-E", that generate colonies of approximately 100 red cells expressing both embryonic and adult globins.[61] These progenitors were found exclusively in the yolk sac at E8 and disappeared by E9. In vitro studies of embryonic stem cell-derived embryoid bodies confirmed that primitive erythroid progenitors, termed EryP-CFC, generate colonies comprised of large nucleated primitive erythroid cells that express both embryonic and adult globins, while BFU-E and CFU-E give rise to colonies comprised of small definitive red cells expressing only adult globins.[62,63] Two overlapping waves of erythroid progenitors were found in differentiating embryoid bodies, an initial primitive erythroid wave followed by a definitive erythroid wave.[63,64]

Studies of hematopoietic progenitor development in the early mouse embryo also revealed two overlapping waves of erythroid progenitors.[65] The first wave consists of EryP-CFC that arise at E7.0 (late primitive streak stage) in the regions of the embryo that give rise to the yolk sac (Fig. 2). EryP-CFC expand in number within the yolk sac over the next 36 hours and then rapidly disapppear. EryP-CFC are associated with macrophage progenitors (Mac-CFC) both in embryonic time and space.[65] Megakaryocyte progenitors (Meg-CFC) have recently been identified in the yolk sacs of unstaged mouse embryos as early as E7.5, indicating that the megakaryocyte lineage also arises during gastrulation.[66,67] Some of these megakaryocyte colonies contained erythroid cells that express both embryonic and adult globins, suggesting that these early megakaryocyte progenitors are associated with the first "primitive" wave of hematopoiesis.

The second wave of erythropoiesis consists of BFU-E that arise in the yolk sac at early somite pair stages (E8.25) prior to the onset of circulation[65] (Fig. 2). Following an initial expansion in the yolk sac, BFU-E enter the bloodstream and expand in numbers within the liver after E10. This definitive wave of erythropoiesis in the yolk sac is associated with macrophage (Mac-CFC), mast cell (Mast-CFC), and granulocyte-macrophage (GM-CFC) progenitors,[65] that also arise in the yolk sac. All of these progenitor types are subsequently found in the bloodstream before being detected in the liver. These kinetics are consistent with the hypothesis that yolk sac-derived hematopoietic progenitors colonize the early fetal liver.

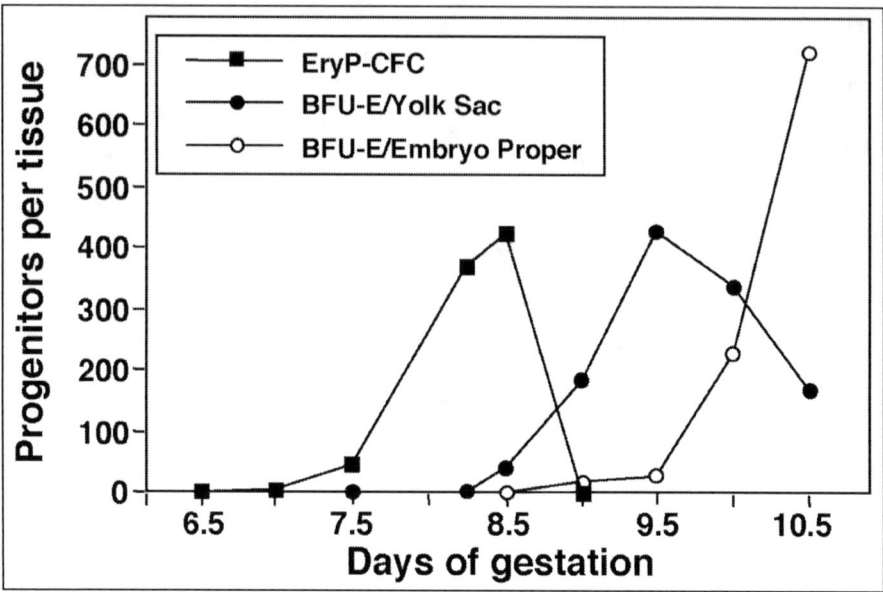

Figure 2. Two waves of erythroid progenitors arise in the gastrulating mouse embryo. The first wave consists of primitive erythroid progenitors that remain confined to the yolk sac (EryP-CFC). The second wave consists of definitive erythroid progenitors (BFU-E) that originate in the yolk sac and are subsequently found within the embryo proper.

Development of Multipotential Myeloid Progenitors

High proliferative potential colony-forming cells (HPP-CFC) are multipotential progenitors that give rise to macroscopic colonies of > 50,000 myeloid cells and represent the earliest multipotential precursor within the hematopoietic hierarchy that can be cultured in vitro without stromal support.[68,69] Because this assay does not rely on adult stroma, it was used to examine the origin of multipotential hematopoietic precursors in the mouse embryo.[70] No HPP-CFC were found at presomite stages, despite examining more than 150 staged embryos. HPP-CFC were first detected at early somite stages (E8.25), exclusively in the yolk sac. Their origin in embryonic time and space coincides with that of BFU-E. The yolk sac remains the predominant site of HPP-CFC expansion (> 100-fold) until the liver begins to serve as the major hematopoietic organ at E11.5.[70] Upon secondary replating, embryonic HPP-CFC give rise to definitive erythroid and macrophage, but not primitive erythroid progenitors. These findings support the hypothesis that HPP-CFC are associated with the definitive wave of hematopoiesis in the yolk sac during late gastrulation (see model, Fig. 3).

Development of Hematopoietic Stem Cells

The development of primitive red cells and myeloid progenitors in the yolk sac of the gastrulating mouse embryo led to the hypothesis that hematopoietic stem cell activity also originates in the yolk sac.[71] However, multiple attempts to engraft adult mouse recipients with yolk sac or fetal liver cells prior to E11.5 were unsuccessful.[40,72-74] Rather, murine hematopoietic stem cells with the ability to repopulate adult recipients were first detected in the aorta-gonad-mesonephros region of the embryo proper at E10.5; one day before their appearance in the liver and yolk sac.[40] However, an "embryonic" hematopoietic stem cell capable of engrafting newborn, but not adult, recipients has been detected as early as E9-9.5 in both the yolk sac and embryo proper of the mouse.[75,76] After being transplanted into newborn recipients,

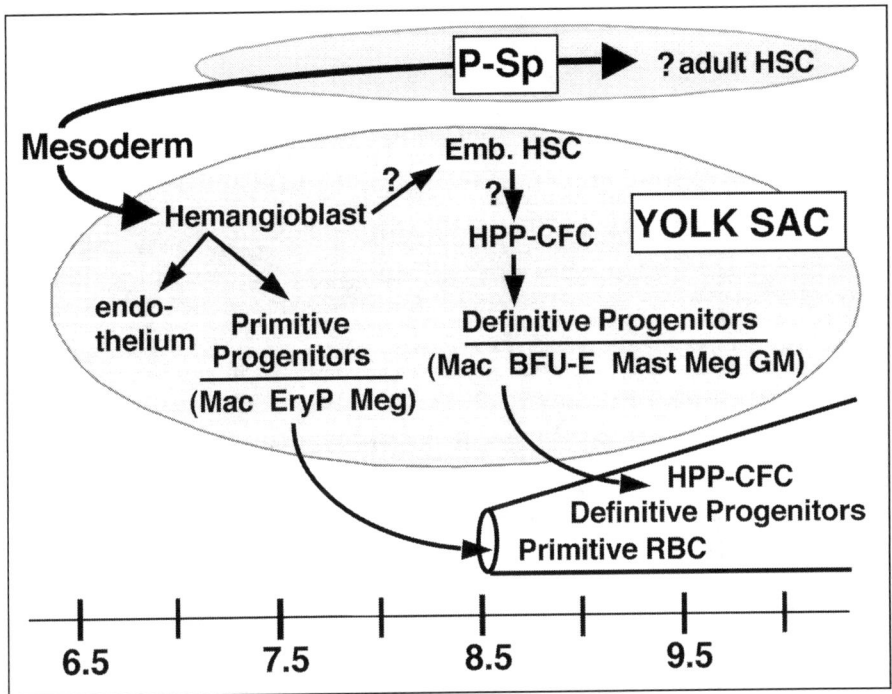

Figure 3. Speculative model of early hematopoietic development in the mouse embryo. Mesoderm cells migrate into yolk sac and embryo proper (P-Sp: para-aortic splanchnopleure). Mesoderm populations migrating into the embryo proper ultimately give rise to adult-repopulating hematopoietic stem cells (adult HSC) at E10.5. Within the yolk sac, mesoderm-derived hemangioblast precursors generate endothelium and two waves of hematopoietic progenitors. The first "primitive" wave contains primitive erythroid (EryP-CFC), macrophage (Mac-CFC) and megakaryocyte (Meg-CFC) progenitors. The second "definitive" wave consists of definitive erythroid (BFU-E), Mac-CFC, Meg-CFC, mast cell (Mast-CFC) and granulocyte-macrophage (GM-CFC). The origin of high proliferative potential colony-forming cells (HPP-CFC) in the yolk sac at E8.25 suggests that it is part of the definitive wave of hematopoiesis that could arise from embryonic hematopoietic stem cells (Emb. HSC) that are capable of engrafting newborn but not adult recipients.

embryonic hematopoietic stem cells can eventually colonize the bone marrow and engraft secondary adult recipients,[75] suggesting that embryonic hematopoietic stem cells undergo a maturation process to become "adult" hematopoietic stem cells. While the developmental origin and the physiologic relevance of embryonic hematopoietic stem cells is not yet known, the appearance of multipotential hematopoietic progenitors (HPP-CFC) in the yolk sac prior to the onset of circulation raises the possibility that embryonic hematopoietic stem cell activity originates in the yolk sac. The developmental origin of hematopoietic stem cells is discussed in greater detail in Chapter 6 (M. Yoder).

Summary: A Model of Yolk Sac Hematopoiesis

The study of unipotential and multipotential hematopoietic progenitors supports a model of blood cell emergence in the mouse embryo as outlined in Figure 3. Mesoderm cells begin to migrate through the primitive streak at E6.5. These mesoderm cells give rise to hemangioblast precursors that in turn generate endothelial cells and two waves of hematopoietic progenitors in the developing yolk sac. The first wave likely consists of primitive erythroid, macrophage, and megakaryocyte progenitors. The transient wave of EryP-CFC becomes a single, nearly synchronous wave of maturing erythroblasts that are necessary for survival and growth of the embryo until at least E12.5, at which time red cells from the fetal liver begin to enter the circulation.

A second wave of definitive erythroid, macrophage, mast cell, and granulocyte/macrophage progenitors subsequently arise in the yolk sac at early somite pair stages (E8.25). These progenitors enter the bloodstream with the onset of circulation and likely go on to colonize the fetal liver. HPP-CFC also arise in the yolk sac at E8.25, indicating that they are associated with the definitive wave of hematopoiesis. The presence of these multipotential precursors in the yolk sac suggests that "embryonic" hematopoietic stem cells, capable of reconstituting newborn but not adult recipients, may originate in the yolk sac. However, transplantation studies with cells from specific regions of precirculation embryos will be required to determine the developmental origin of hematopoietic stem cell activity in the mammalian embryo.

Acknowledgements

I wish to thank my colleagues in the laboratory, particularly Paul Kingsley and Kathleen McGrath, as well as collaborators, Gordon Keller and Merv Yoder, for their wisdom and friendship. This research has been supported by the National Institutes of Health, the American Cancer Society, and the University of Rochester Strong Children's Research Center.

References

1. Copp AJ. Death before birth: Clues from gene knockouts and mutations. Trends Genet 1995; 11:87-93.
2. Gardner RL, Rossant J. Investigation of the fate of 4.5 day post coitum mouse inner cell mass cells by blastocyst injection. J Emb Exp Morph 1979; 52:141-152.
3. Downs KM, Gifford S, Blahnik M et al. Vascularization in the murine allantois occurs by vasculogenesis without accompanying erythropoiesis. Development 1998; 125:4507-4520.
4. Cross JC, Werb Z, Fisher SJ. Implantation and the placenta: Key pieces of the developmental puzzle. Science 1994; 266:1508-1518.
5. Lawson KA, Meneses JJ, Pedersen RA. Clonal analysis of epiblast fate during germ layer formation in the mouse embryo. Development 1991; 113.891-911.
6. Tam P, Williams A, Chan WY. Gastrulation in the mouse embryo: Ultrastructural and molecular aspects of germ layer morphogenesis. Micro Res Techn 1993; 26:301-328.
7. Jollie WP. Development, morphology, and function of the yolk-sac placenta of laboratory rodents. Teratol 1990; 41:361-381.
8. Wilt FH. Erythropoiesis in the chick embryo: The role of endoderm. Science 1965; 147:1588-1590.
9. Miura Y, Wilt FH. Tissue interaction and the formation of the first erythroblasts of the chick embryo. Dev Biol 1969; 19:201-211.
10. Palis J, McGrath KE, Kingsley PD. Initiation of hematopoiesis and vasculogenesis in murine yolk sac explants. Blood 1995; 86:156-163.
11. Bielinska M, Narita N, Heikinheimo M et al. Erythropoiesis and vasculogenesis in embryoid bodies lacking visceral yolk sac endoderm. Blood 1996; 88:3720-3730.
12. Belaousoff M, Farrington SM, Baron MH. Hematopoietic induction and respecification of A-P identity by visceral endoderm signaling in the mouse embryo. Development 1999; 125:5009-5018.
13. Dyer MA, Farrington SM, Mohn D et al. Indian hedgehog activates hematopoiesis and vasculogenesis and can respecify neurectodermal cell fate in the mouse embryo. Development 2001; 128:1717-1730.
14. Byrd N, Becker S, Maye P et al. Hedgehog is required for murine yolk sac angiogenesis. Development 2002; 129:361-372.
15. Winnier G, Blessing M, Labosky PA et al. Bone morphogenetic protein-4 is required for mesoderm formation and patterning in the mouse. Genes Dev 1995; 9:2105-2116.
16. Johansson BM, Wiles MV. Evidence for involvement of activin A and bone morphogenetic protein 4 in mammalian mesoderm and hematopoietic development. Mol Cell Biol 1995; 15:141-151.

17. Maximow AA. Untersuchungen uber blut und bindegewebe 1. Die fruhesten entwicklungsstadien der blut und bindegewebszellan bein saugeberembryo, bis zum anfang der blutbilding unden leber. Arch Mikroskop Anat 1909; 73:444-561.

18. Sabin FR. Studies on the origin of blood vessels and red blood corpuscles as seen in the living blastoderm of chicks during the second day of incubation. Contrib Embryol 1920; 9:213-262.

19. Haar JL, Ackerman GA. A phase and electron microscopic study of vasculogenesis and erythropoiesis in the yolk sac of the mouse. Anat Rec 1971; 170:199-224.

20. Sasaki K, Matsamura G. Haemopoietic cells of yolk sac and liver in the mouse embryo: A light and electron microscopical study. J Anat 1986; 148:87-97.

21. Sorenson GD. An electron microscopic study of hematopoiesis in the yolk sac. Lab Invest 1961; 10:178-193.

22. Smith RA, Glomski CA. Embryonic and fetal hemopoiesis in the Mongolian gerbil (Meriones unguiculatus). Anat Rec 1977; 189:499-517.

23. Tiedemann K. On the yolk sac of the cat. Cell Tiss Res 1977; 183:171-189.

24. Takashina T. Haemopoiesis in the human yolk sac. J Anat 1987; 151:125-135.

25. Risau W, Lemmon V. Changes in the vascular extracellular matrix during embryonic vasculogenesis and angiogenesis. Dev Biol 1988; 125:441-450.

26. Drake CJ, Fleming PA. Vasculogenesis in the day 6.5 to 9.5 mouse embryo. Blood 2000; 95:1671-1679.

27. Murray PDF. The development in vitro of the blood of the early chick embryo. Proc Royal Soc London 1932; 111:497-521.

28. Kallianpur AR, Jordan JE, Brandt SJ. The SCL/TAL-1 gene is expressed in progenitors of both the hematopoietic and vascular systems during embryogenesis. Blood 1994; 83:1200-1208.

29. Anagnostou A, Liu Z, Steiner M et al. Erythropoietin receptor mRNA expression in human endothelial cells. Proc Natl Acad Sci USA 1994; 91:3974-3978.

30. Hashiyama M, Iwama A, Ohshiro K et al. Predominant expression of a receptor tyrosine kinase, tie, in hematopoietic stem cells and B cells. Blood 1996; 87:93-101.

31. Watt SM, Gschmeissner SE, Bates PA. PECAM-1: Its expression and function as a cell adhesion molecule on hemopoietic and endothelial cells. Leuk Lymphoma 1995; 17:229-244.

32. Shalaby F, Rossant J, Yamaguchi TP et al. Failure of blood-island formation and vasculogenesis in Flk-1-deficient mice. Nature 1995; 376(6535):62-6.

33. Schuh AC, Faloon P, Hu Q-L et al. In vitro hematopoietic and endothelial potential of flk-/- embryonic stem cells and embryos. Proc Natl Acad Sci USA 1999; 96:2159-2164.

34. Shivdasani RA, Mayer EL, Orkin SH. Absence of blood formation in mice lacking T-cell leukemia oncoprotein tal-1/SCL. Nature 1995; 373:432-434.

35. Robb L, Lyons I, Li R et al. Absence of yolk sac hematopoiesis from mice with a targeted disruption of the scl gene. Proc Natl Acad Sci USA 1995; 92:7075-7079.

36. Warren AJ, Colledge WH, Carlton MBL et al. The oncogenic cysteine-rich LIM domain protein is essential for erythroid development. Cell 1994; 78:45-57.

37. Visvader JE, Fujiwara Y, Orkin SH. Unsuspected role for the T-cell leukemia protein SCL/tal-1 in vascular development. Genes Dev 1998; 12(4):473-479.

38. Yamada Y, Pannell R, Forster A et al. The oncogenic LIM-only transcription factor Lmo2 regulates angiogenesis but not vasculogenesis in mice. Proc Natl Acad Sci USA 2000; 97:320-324.

39. Choi K, Kennedy M, Kazarov A et al. A common precursor for hematopoietic and endothelial cells. Development 1998; 125:725-732.

40. Falloon PE, Arentson A, Kazarov A et al. Basic fibroblast growth factor positively regulates hematopoietic development. Development 2000; 127:1931-1941.

41. Robertson SM, Kennedy M, Shannon JM et al. A transitional stage in the commitment of mesoderm to hematopoiesis requiring the transcription factor SCL/tal-1. Development 2000; 127:2447-2459.

42. Nishikawa S-I, Nishikawa S, Kawamoto H et al. In vitro generation of lymphohematopoietic cells from endothelial cells purified from murine embryos. Immunity 1998; 8:761-769.

43. Yamashita J, Itoh H, Hirashima M et al. Flk1-positive cells derived from embryonic stem cells serve as vascular progenitors. Nature 2000; 408:92-96.

44. Huber TL, Kouskoff V, Fehling HJ et al. Haemangioblast commitment is initiated in the primitive streak of the mouse embryo. Nature 2004; 432:625-630.
45. Smith RA, Glomski CA. "Hemogenic endothelium" of the embryonic aorta: Does it exist? Dev Comp Immunol 1982; 6:359-368.
46. Muller AM, Medvinsky A, Strouboulis J et al. Development of hematopoietic stem cell activity in the mouse embryo. Immunity 1994; 1:291-301.
47. de Bruijn MFTR, Speck NA, Peeters MCE et al. Definitive hematopoietic stem cells first develop within the major arterial regions of the mouse embryo. EMBO J 2000; 19:2465-2474.
48. Okuda T, van Deursen J, Hiebert SW et al. AML-1, the target of multiple chromosomal translocations in human leukemia, is essential for normal fetal liver hematopoiesis. Cell 1996; 84:321-330.
49. Wang Q, Stacy T, Binder M et al. Disruption of the Cbfa2 gene causes necrosis and hemorrhaging in the central nervous system and blocks definitive hematopoiesis. Proc Natl Acad Sci USA 1996; 93:3444-3449.
50. North T, Gu T-L, Stacey T et al. Cbfa is required for the formation of intra-aortic hematopoietic clusters. Development 1999; 126:2563-2575.
51. Leder A, Kuo A, Shen M et al. In situ hybridization reveals coexpression of embryonic and adult α globin genes in the earliest murine erythrocyte progenitors. Development 1992; 116:1041-1049.
52. Silver L, Palis J. Initiation of murine embryonic erythropoiesis: A spatial analysis. Blood 1997; 89:1154-1164.
53. McGrath KE, Koniski AD, Malik J et al. Circulation is established in a step-wise pattern in the mammalian embryo. Blood 2003; 101:1669-1676.
54. Ji RP, Phoon CKL, Aristizábal O et al. Onset of cardiac function during early mouse embryogenesis coincides with entry of primitive erythroblasts into the embryo proper. Circ Res 2003; 92:133-135.
55. Fantoni A, de la Chapelle A, Rifkind RA et al. Erythroid cell development in fetal mice: Synthetic capacity for different proteins. J Mol Biol 1968; 33:79-91.
56. Steiner R, Vogel H. On the kinetics of erythroid cell differentiation in fetal mice: I. Microspectrophotometric determination of the hemoglobin content in erythroid cells during gestation. J Cell Physiol 1973; 81:323-338.
57. Sangiorgi F, Woods CM, Lazarides E. Vimentin downregulation is an inherent feature of murine erythropoiesis and occurs independently of lineage. Development 1990; 110:85-96.
58. Kingsley PD, Malik J, Fantauzzo KA et al. Yolk sac-derived primitive erythroblasts enucleate during mammalian embryogenesis. Blood 2004; 104:19-25.
59. Iscove NN, Sieber F. Erythroid progenitors in mouse bone marrow detected by macroscopic colony formation in culture. Exp Hematol 1975; 3:32-43.
60. Stephenson JR, Axelrad A, McLeod D et al. Induction of colonies of hemoglobin-synthesizing cells by erythropoietin in vitro. Proc Natl Acad Sci USA 1971; 68:1542-1546.
61. Wong PMC, Chung SW, Chui DHK et al. Properties of the earliest clonogenic hematopoietic precursors to appear in the developing murine yolk sac. Proc Natl Acad Sci USA 1986; 83:3851-3854.
62. Keller G, Kennedy M, Papayannopoulou T et al. Hematopoietic commitment during embryonic stem cell differentiation in culture. Mol Cell Biol 1993; 13:473-486.
63. Kennedy M, Firpo M, Choi K et al. Identification of a common precursor for primitive and definitive hematopoiesis. Nature 1997; 386:488-493.
64. Nakano T, Kodama H, Honjo T. In vitro development of primitive and definitive erythrocytes from different precursors. Science 1996; 272:722-724.
65. Palis J, Robertson S, Kennedy M et al. Development of erythroid and myeloid progenitors in the yolk sac and embryo proper of the mouse. Development 1999; 126:5073-5084.
66. Xu M-J, Matsuoka S, Yang F-C et al. Evidence for the presence of primitive megakaryocytopoiesis in the early yolk sac. Blood 2001; 97:2016-2022.
67. Xie X, Chan RJ, Johnson SA et al. Thrombopoietin promotes mixed lineage and megakaryocytic colony forming cell growth but inhibits primitive and definitive erythropoiesis in cells isolated from early murine yolk sacs. Blood 2003; 101:1329-1335.
68. Bradley TR, Hodgson GS. Detection of primitive macrophage progenitor cells in mouse bone marrow. Blood 1979; 54:1446-1450.

69. Bertoncello I. Status of high proliferative potential colony-forming cells in the hematopoietic stem cell hierarchy. Curr Top Microbiol Immunol 1992; 177:83-94.

70. Palis J, Chan RJ, Koniski A et al. Spatial and temporal emergence of high proliferative potential hematopoietic precursors during murine embryogenesis. Proc Natl Acad Sci USA 2001; 98:4528-453.

71. Metcalf D, Moore M. Haemopoietic cells. London: North-Holland Publishing Company, 1971.

72. Harrison DE, Astle CM, DeLaittre JA. Processing by the thymus is not required for cells that cure and populate W/W^v recipients. Blood 1979; 54:1152-1157.

73. Sonoda T, Hayashi C, Kitamura Y. Presence of mast cell precursors in the yolk sac of mice. Dev Biol 1983; 97:89-94.

74. Ema H, Nakauchi H. Expansion of hematopoietic stem cells in the developing liver of the mouse embryo. Blood 2000; 95:2284-2288.

75. Yoder MC, Hiatt K, Dutt P et al. Characterization of definitive lymphohematopoietic stem cells in the day 9 murine yolk sac. Immunity 1997; 7:335-344.

76. Yoder MC, Hiatt K, Mukherjee P. In vivo repopulating hematopoietic stem cells are present in murine yolk sac at day 9.0 postcoitus. Proc Natl Acad Sci USA 1997; 94:6776-6780.

CHAPTER 6

Long-Term Reconstituting Hematopoietic Stem Cell Capacity in the Embryo

Mervin C. Yoder

Introduction

Hematopoiesis is the dynamic process whereby all formed elements of the blood arise from multipotent precursor cells. In the adult mouse, hematopoiesis occurs primarily within the medullary cavity of bone and in the spleen. During murine development, however, hematopoiesis is found in numerous sites within and outside the embryo proper.[1] The first blood cells in the mouse embryo arise between embryonic day 7 and 8 (E7-8) in the yolk sac.[2,3] Primitive erythroid progenitor production is restricted to the yolk sac and while these cells first differentiate into mature nucleated primitive red blood cells expressing embryonic globins, recent evidence indicates that eventually these cells enucleate intravascularly.[4] Other recent evidence indicates that "definitive" erythroid and myeloid progenitors, including high proliferative potential colony-forming cells, are also produced in the yolk sac as early as E8.25.[5,6]

Hematopoiesis, including hematopoietic stem cell expansion, is sustained in the yolk sac until approximately E12-13 when the fetal liver has become the predominant hematopoietic organ.[7-9] Two other sites of hematopoietic stem and progenitor cell development, the aorta-gonad-mesonephros (AGM) region and placenta, emerge as important sites for stem cell development prior to initiation of liver hematopoietic activity.[9-12] The liver displays sustained hematopoietic activity throughout the remainder of in utero development and continues to serve as a hematopoietic site for several weeks postnatally.[13-15] Hematopoiesis emerges in the murine spleen and bone marrow late in mouse gestation but is well developed in both sites by E18.[16] Several lines of evidence confirm that the bone marrow is seeded by circulating HSC derived from the rodent fetal liver and that these HSC serve as the source of the marrow HSC pool for the adult host.[17,18]

This brief overview provides some of the evidence that hematopoiesis is a developmentally regulated and tissue-specific process. The stem cell theory of hematopoiesis predicts that initiation of blood cell production in these various tissues during murine development should be preceded by the appearance of hematopoietic stem cells (HSC).[19] In this chapter, we will review some evidence for the emergence of long-term reconstituting HSC activity in the early murine embryo.

Hematopoietic Stem Cell Development, edited by Isabelle Godin and Ana Cumano.
©2006 Eurekah.com and Kluwer Academic / Plenum Publishers.

General Principles of Hematopoietic Stem Cell Emergence during Ontogeny

Moore and Owen[20] proposed that the embryonic murine liver depends on an inflow of circulating HSC for the initiation of hematopoiesis during embryonic development. Early in vitro and in vivo studies in the mouse supported the apparent requirement of an exogenous source of hematopoietic stem cells to initiate fetal liver hematopoiesis.[8,21] Removal of the liver prior to the 28 somite pair stage (E9.5) and grafting of the liver primordium into conditioned adult recipients did not result in the growth of any hematopoietic cells within the graft unless the conditioned recipients were also transplanted with hematopoietic cells. In the presence of the transplanted hematopoietic cells, multilineage hematopoiesis was observed in the fetal liver grafts. Likewise, when E7 embryos were dissected free of their yolk sac membranes and cultured for 2 days in vitro, embryonic growth and development appeared normal, however no blood cells or hematopoietic colony forming cells developed in any portion of the embryo including the liver.[22] If the liver is dependent on an exogenous source of HSC, as these data suggest, the primary unanswered question becomes finding the source of the circulating HSC that seed the liver primordium.

The most direct assay of HSC activity is transplantation of test cells into a myeloablated host and subsequent identification of donor-derived cellular reconstitution of host blood with lymphoid and myeloid cells.[23] Sustained production of donor-derived blood cells for more than 4 months constitutes evidence of long-term repopulating ability of the HSC.[23] Efforts to directly assay for the presence of HSC activity in the yolk sac via transplantation of yolk sac or embryo proper cells into recipient animals have yielded discrepant results that appear to correlate to some extent with the age of recipient animal used.

An additional variable in defining the spatial and temporal emergence of HSC has been to attempt to isolate hematopoietic cells from the yolk sac prior to establishment of the systemic circulation. After the blood is circulating throughout the embryo, it becomes nearly impossible with standard dissection techniques to be sure of the site of origin of cells within a tissue. While the heart is known to begin beating at the 4-8 somite (E8.25) stage of development, the distribution of red blood cells in the arterial and venous circulations does not become fully equilibrated until 26-30 somite pairs (E10.0) just prior to seeding of the developing liver with hematopoietic cells.[24] Thus, movement of yolk sac derived cells into the systemic circulation does not occur abruptly but gradually. In fact, hematopoietic progenitor cells remain enriched in the yolk sac circulation as late as E10.5, suggesting that this organ continues to serve as a site of progenitor cell emergence or adhesion even after the systemic vascular system is intact and functional.[24]

Emergence of Hematopoietic Stem Cells in the Yolk Sac

Evidence has been published that the yolk sac is a site for HSC emergence. Transplantation of E8 yolk sac cells into the yolk sacs of congenic nonablated mice in utero indicated that HSC activity was present.[25] Donor yolk sac cells gave rise to both myeloid and lymphoid cells in recipient mice that survived and were tested weeks after birth. Toles et al[26] reported that transplantation of E9 yolk sac cells into c-kit deficient congenic fetal mice in utero via a transplacental route resulted in donor type reconstitution of the erythroid lineage for more than 5 months. A total of 14 dams were also reconstituted with the E9 yolk sac or peripheral blood cells. Engraftment of yolk sac cells in the pregnant dams is most interesting as numerous investigators have been unable to identify donor type hematopoietic elements when yolk sac cells prior to E11 are transplanted into c-Kit deficient or lethally irradiated adult recipient congenic mice.[27-30] Thus, yolk sac cells prior to E11 appear to contain HSC if injected into embryonic recipients (and their mothers) in utero but repopulating ability cannot be demonstrated using these same cells upon direct transplantation into myeloablated adult recipients.

There may be several explanations for the failure of yolk sac cells to engraft in the adult recipient mice including: (1) yolk sac hematopoietic progenitors do not possess intrinsic HSC activity, (2) yolk sac HSC are unable to engraft in the adult hematopoietic microenvironment, (3) the frequency of yolk sac HSC activity is below detection using standard transplantation protocols, and (4) yolk sac hematopoietic progenitors demonstrate low expression of major histocompatibility class I antigens (MHC I) and may be removed by adult host-derived natural killer cells upon transplantation.

The inability of donor yolk sac cells prior to E11 to engraft adult hosts may be due to their inability to home and engraft in the marrow compartment of adult mice. Since the liver continues to function as a hematopoietic organ for several weeks after birth, we speculated that yolk sac HSC prior to E11 may engraft in conditioned newborn mice. An intrapartum conditioning regimen of subcutaneous busulfan injected into pregnant dams on day 18 of gestation yielded newborn pups in a sublethally myeloablated state.[31] Injection of these conditioned newborn pups with yolk sac cells obtained from congenic mice at E10 resulted in engraftment in erythroid, myeloid, and lymphoid lineages in primary and secondary recipient mice.[32] The level of donor cell engraftment was correlated with the concentration of donor cells transplanted. In a subsequent study, E9 yolk sac cells were demonstrated to contain hematopoietic stem cells with long-term repopulating ability upon intravenous or intrahepatic injection into sublethally conditioned newborn mice.[33] Engraftment in primary recipients up to 11 months posttransplant and secondary recipients up to 6 months posttransplant was documented to be multilineage. Of interest, while the yolk sac cells were injected into the newborn recipients for the primary transplants, bone marrow cells harvested from these primary hosts at 11 months contained yolk sac-derived HSC activity in the bone marrow compartment capable of repopulating all lineages in the secondary lethally irradiated adult recipients. These results suggest that yolk sac HSC prior to E11 are unable to directly repopulate conditioned adult recipient mice, however, once engrafted in conditioned newborn recipients, the yolk sac-derived hematopoietic stem cells apparently migrate to the marrow cavity and function long-term.

The hematopoietic repopulating ability and clonogenic behavior of E9 yolk sac hematopoietic cells expressing the c-Kit and CD34 cell surface molecules were also tested as these antigens have been used to isolate HSC from fetal and adult mice.[34-36] The percentage of cells expressing CD34 and the level of CD34 expression was highest in the yolk sac at E9 and decreased throughout murine ontogeny so that bone marrow cells express low levels of CD34.[30] Transplantation of 1000 CD34[+] E9 yolk sac cells into conditioned newborn recipients resulted in long-term multilineage engraftment and yolk sac-derived repopulating activity from the marrow of the primary recipients engrafted secondary recipients long-term in myeloid and lymphoid lineages. E9 yolk sac cells expressing CD34 and c-Kit were highly enriched for high proliferative potential colony forming cells. Transplantation of CD34[+]c-Kit[+]E9 yolk sac cells into conditioned newborn mice demonstrated that the yolk sac cells gave rise to multilineage repopulating ability. While the precise mechanisms remain elusive, it is clear that the yolk sac is composed of hematopoietic cells with repopulating ability that engraft in newborn mice and remain active in the sustained production of lymphomyelopoiesis long term in aging adult mice.

Failure of yolk sac cells to engraft in adult conditioned mice may also be related to limiting numbers of stem cells in the yolk sac that upon transplantation in the adult are insufficient in number to outcompete the host cells. In an effort to isolate yolk sac cells prior to the establishment of the circulation in the embryo and yet to permit sufficient growth and expansion of cells for accurate assessment of in vivo engraftment, Cumano and coworkers[29,37] developed in vitro organotypic culture conditions for yolk sac hematopoietic cell expansion.

This system tests the capacity of the explanted tissue to expand the HSC in situ and is assayed as successful LTR-HSC function in myeloablated adult recipient mice. Explanted yolk sac tissue was able to give rise to myeloid progenitor cells and to reconstitute the myeloid lineage in lethally irradiated hosts for up to 3 months. However, under the culture conditions employed, yolk sac cells derived from 8-12 somite-pair staged embryos, cultured for up to 4 days failed to contribute to the lymphoid lineage in recipient mice. These data suggest that precirculation yolk sac tissue is devoid of the capacity to support the expansion of HSC that can repopulate lymphomyelopoiesis in lethally irradiated adult mice. Furthermore, while yolk sac cells in organotypic cultures proliferated, MHC I expression remained low, and these cells remained at risk for elimination by host NK cells in the adult recipient mice.[29] Transplantation of 4 day organotypically cultured precirculation yolk sac cells into sublethally myeloablated immuno-deficient mice (which are NK cell deficient and thus less able to eliminate the donor yolk sac cells that have low MHC I expression) also failed to reveal evidence of LTR-HSC. These data suggest that fresh or cultured yolk sac cells do not have the capacity to engraft in adult mice and therefore are not a site for emergence of HSC.

A completely different conclusion can be drawn from the data presented by Matsuoka et al[38] and Kyba et al.[39] Precirculation yolk sac cells (E8.0) were dissected free of the embryo and cocultured with an endothelial cell line originally derived from the aorta-gonad-mesonephros (AGM) region of the murine embryo. This cell line, AGM-S3 had previously been reported to support the in vitro proliferation of primitive murine hematopoietic progenitor cells including HSC.[40] While the freshly isolated precirculation yolk sac cells failed to engraft in lethally irradiated adult recipients, ex vivo coculture of the yolk sac cells for 4 days with the AGM-S3 cell line resulted in multilineage long-term engraftment. Likewise, transduction of yolk sac cells with HoxB4 (a homeotic transcription factor) and coculture on a stromal cell monolayer induced the appearance of competence to engraft in primary and secondary adult conditioned mice.[39] These results suggest that HSC activity is present in the yolk sac prior to establishment of the systemic circulation, but that coculture with the AGM cell line or overexpression of the HoxB4 protein is necessary and sufficient for maturing the cells for engraftment in the conditioned adult mice.

An overall summation of the above data suggests that < E11.0 yolk sac HSC are detectable upon transplantation into embryonic or newborn but not adult recipients, unless the yolk sac cells are first cocultured under specific conditions. An interesting recent report suggests that while 11 day post-coitus (dpc) yolk sacs fail to expand or support HSC, this activity is clearly present at 12 dpc. These results suggest that early YS HSC may not need exposure to AGM cells but can mature in situ in the 12 dpc yolk sac into liver repopulating cells or alternatively, that HSC derived from the AGM enter the circulation and seed the YS.[9] The recent identification of the placenta as a site of HSC development raises further questions as to whether the HSC in this organ emerge directly from placental precursors or are seeded from other sites such as the AGM or YS.

Hematopoietic Stem Cell Ontogeny in the Embryo Proper

Hematopoiesis in the early embryo proper emerges from a region surrounding the developing aorta. The aorta develops in close association with the mesonephros, mesentery, and gonads in the splanchnopleura formed by the interaction of mesoderm and endoderm cells. This region of the embryo has been referred to as the para-aortic splanchnopleura (P-Sp) prior to E10 and the AGM region thereafter.[41,42]

When the P-Sp and AGM regions were isolated from the embryo and the tissue directly transplanted into lethally irradiated adult mice, HSC activity was first detectable in 10 dpc cells from the AGM region.[10,28] These studies further demonstrated that HSC activity was concomitantly present in yolk sac, liver, and AGM region at 11 dpc, suggesting that the AGM was the first site of HSC appearance in the early embryo.[10] If the P-Sp region was dissected free of yolk sac cells prior to the establishment of the circulation (8 dpc) and cultured in vitro, HSC activity appeared autonomously in the P-Sp cultured tissue.[29] The HSC stem cell activity measured in the adult recipient mice was noted to be multilineage and maintained long term. HSC activity was not present in similarly cultured yolk sac explants, as noted above.[29] These data suggest that the first and only site of HSC emergence occurs in the P-Sp/AGM region of the embryo proper.[29]

P-Sp cells isolated before 10 dpc do not demonstrate HSC with repopulating activity in conditioned adult mice, but can acquire HSC activity if cultured in vitro. In contrast, 9 dpc P-Sp cells are capable of directly engrafting and repopulating all blood lineages in conditioned newborn recipient mice.[30] Thus, similar to the paradigm displayed by yolk sac cells, there are hematopoietic cells that emerge in the P-Sp that do not immediately possess the capacity to engraft in conditioned adult mice. Unlike the yolk sac environment, however, the P-Sp/AGM region is sufficient to induce these hematopoietic cells to "mature" into HSC that engraft and repopulate conditioned adult mice.

Hematopoietic cells within the P-Sp/AGM are concentrated as a cluster along the ventral floor of the aorta and in the underlying surrounding mesenchyme.[43,44] Similar clusters of hematopoietic cells have been identified in numerous species during embryogenesis.[19,45] HSC have also been identified in clusters along the floor of the vitelline and umbilical arteries in the mouse embryo.[45] Some evidence points to the generation and maturation of hematopoietic cells beneath the aortic endothelium in the mesenchyme and migration of the HSC through the endothelium to form the clusters.[45-47] Other work supports the emergence of hematopoietic cells directly from hemogenic endothelium or hemangioblasts in the AGM region.[48-51]

While the AGM region is a site for HSC emergence, this region poorly supports hematopoietic progenitor cell differentiation. When P-Sp and AGM tissue is dissected and a single cell suspension prepared and plated in hematopoietic colony forming cell assays, few committed erythroid and myeloid progenitors are detectable at any stage of ontogeny.[46,52,53] These data suggest that the P-Sp/AGM region is a unique hematopoietic environment that supports HSC formation but lacks ability to support significant differentiation of the HSC into committed progenitor progeny. Thus, this tissue has been proposed to be the source of HSC that migrate to the liver and establish the HSC pool that sustains hematopoiesis life long.[46,47]

Summary

At present, none of the data presented can resolve the controversy as to the site of origin of the HSC that migrate to and seed the liver to establish definitive hematopoiesis. There are no published data that have directly identified the first hematopoietic cells to seed the liver. All interpretations to this point are circumstantial. A summary of three viable hypotheses is depicted in Figure 1. Our interpretation of the published literature is reconciled most closely with the hypothesis that mesoderm is allocated to both the yolk sac and P-Sp regions during development. In these distinctly separate sites, HSC are formed. While yolk sac HSC do not acquire the capacity to engraft in adult recipients while retained within the yolk sac microenvironment, these cells can display such activity upon further "maturation". In the P-Sp region, HSC emerge but await further "maturation" within the AGM microenvironment before displaying the ability to engraft in conditioned adult recipient mice. Further studies which track the fates of mesoderm cells during these early stages of embryonic development may be required to address the intriguing question of the origin of the HSC which populate the fetal liver.

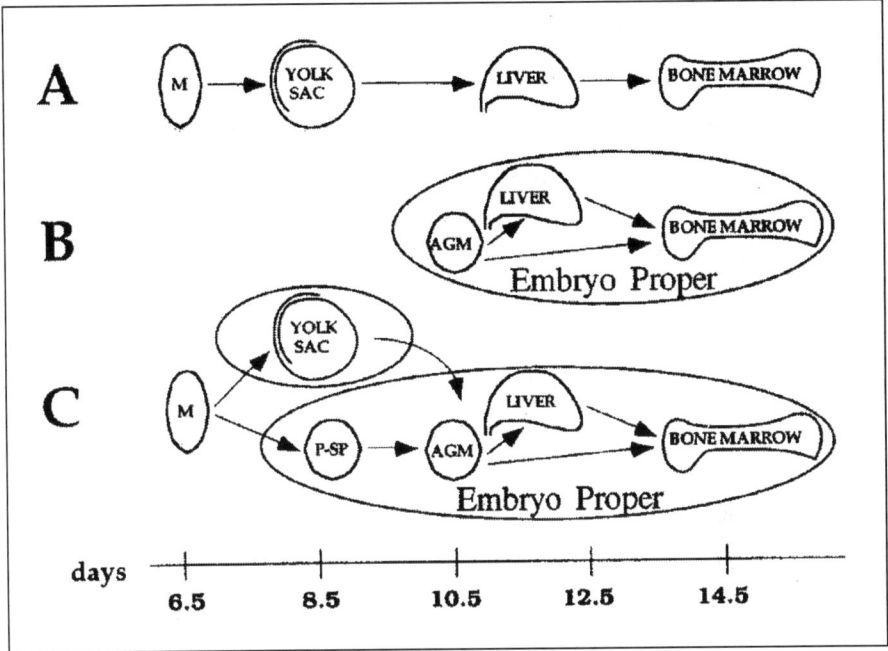

Figure 1. Models of hematopoietic stem cell (HSC) ontogeny. A) Single stem cell model with HSCs arising in the yolk sac from mesoderm. B) Single stem cell model with HSCs arising within the embryo proper, specifically the aorto-gonad-mesonephros (AGM) region. C) Composite stem cell model whereby mesoderm cells are allocated to the yolk sac and the paraaortic-splanchnopleure (P-Sp) region. HSC emerging from these sites further mature within the AGM or liver to gain competence to engraft in adult mice as definitive HSC. Reprinted with permission from Zon LI, ed. Hematopoiesis: A developmental approach. New York: Oxford Univ. Press 2001:188.

References

1. Keller G, Lacaud G, Robertson S. Development of the hematopoietic system in the mouse. Exp Hematol 1999; 27:777-787.
2. Barker J. Development of the mouse hematopoietic system: I. Types of hemoglobin produced in embryonic yolk sac and liver. Dev Biol 1968; 18:14-29.
3. Haar J, Ackerman G. A phase and electron microscopic study of vasculogenesis and erythropoiesis in the yolk sac of the mouse. Anat Rec 1970; 170:199-224.
4. Kingsley P, Malik J, Fantauzzo K et al. Yolk-sac-derived primitive erythroblasts enucleate during mammalian embryogenesis. Blood 2004; 104:19-25.
5. Palis J, Chan RJ, Koniski A et al. Spatial and temporal emergence of high proliferative potential hematopoietic precursors during murine embryogenesis. Proc Nat Acad Sci USA 2001; 98:4528-4533.
6. Cumano A, Furlonger C, Paige C. Differentiation and characterization of B-cell precursors detected in the yolk sac and embryo body of embryos beginning at the 10- to 12-somite stage. Proc Natl Acad Sci USA 1993; 90:6429-6433.
7. Rifkind R, Chui D, Epler H. An ultrastructural study of early morphogenetic events during the establishment of fetal hepatic erythropoiesis. J Cell Biol 1969; 40:343-365.
8. Johnson G, Moore M. Role of stem cell migration in initiation of mouse fetal liver hematopoiesis. Nature 1975; 258:726-728.

9. Kumaravelu P, Hook L, Morrison A et al. Quantitative developmental anatomy of definitive haematopoeitic stem cells/long-term repopulating units (HSC/RUs): Role of the aorta-gonad-mesonephros (AGM) region and the yolk sac in colonisation of the mouse embryonic liver. Development 2002; 129:4891-4899.

10. Muller A, Medvinsky A, Strouboulis J et al. Development of hematopoietic stem cell activity in the mouse embryo. Immunity 1994; 1:291-301.

11. Ottersbach K, Dzierzak E. The murine placenta contains hematopoietic stem cells within the vascular labyrinth region. Dev Cell 2005; 8:377-387.

12. Gekas C, Dieterlen-Lievere F, Orkin S et al. The placenta is a niche for hematopoietic stem cells. Dev Cell 2005; 8:365-375.

13. Wolf NS, Bertoncello I, Jiang D et al. Developmental hematopoiesis from prenatal to young-adult life in the mouse model. Exper Hematol 1995; 23:142-146.

14. Harrison D, Zhong R, Jordan C et al. Relative to adult marrow, fetal liver repopulates nearly five times more effectively long-tem than short-term. Exper Hematol 1997; 25:293-297.

15. Kurata H, Mancini G, Alespeiti G et al. Stem cell factor induces proliferation and differentiation of fetal progenitor cells in the mouse. Br J Haematol 1998; 101:676-687.

16. Grossi C, Velardi A, Cooper M. Postnatal liver hemopoiesis in mice: Generation of preB cells, granulocytes, and erythrocytes in discrete colonies. J Immunol 1985; 135:2303-2311.

17. Clapp D, Freie B, Lee W-H et al. Molecular evidence that in situ-transduced fetal liver hematopoietic stem/progenitor cells give rise to medullary hematopoiesis in adult rats. Blood 1995; 86:2113-2122.

18. Gothert J, Gustin S, Hall M et al. In vivo fate-tracing studies using the SCl stem cell enhancer: Embryonic hematopoietic stem cells significantly contribute to adult hematopoiesis. Blood 2005; 105:2724-2732.

19. Zon L. Developmental biology of hematopoiesis. Blood 1995; 86:2876-2891.

20. Moore M, Owen J. Stem-cell migration in developing myeloid and lymphoid systems. The Lancet 1967; 658-659.

21. Houssaint E. Differentiation of the mouse hepatic primordium. II. Extrinsic origin of the haemopoietic cell line. Cell Diff 1981; 10:243-247.

22. Moore M, Metcalf D. Ontogeny of the haemopoietic system: Yolk sac origin of in vivo and in vitro colony forming cells in the developing mouse embryo. Br J Haematol 1970; 18:279-296.

23. Orlic D, Bodine D. What defines a pluripotent hematopoietic stem cell (PHSC): Will the real PHSC please stand up! Blood 1994; 84:3991-3994.

24. McGrath K, Koniski A, Malik J et al. Circulation is established in a stepwise pattern in the mammalian embryo. Blood 2003; 101:1669-1676.

25. Weissman I, Papaioannou V, Gardner D. Fetal hematopoietic origins of the adult hematolymphoid system. In: Clarkson B, Marks PA, Till JE, eds. Differentiation of Normal and Neoplastic Hematopoietic Cells. Cold Spring Harbor, NY: Cold Spring Harbor Laboratory, 1978:33-47.

26. Toles JF, Chui DH, Belbeck LW et al. Hematopoietic stem cells in murine embryonic yolk sac and peripheral blood. Proc Natl Acad Sci USA 1990; 86:7456-7459.

27. Harrison D, Astle C, DeLaittre J. Processing by the Thymus is not required for cells that cure and populate W/Wv recipients. Blood 1979; 54:1152-1157.

28. Medvinsky A, Dzierzak E. Definitive hematopoiesis is autonomously initiated by the AGM region. Cell 1996; 86:897-906.

29. Cumano A, Ferraz JC, Klaine M et al. Intraembryonic, but not yolk sac hematopoietic precursors, isolated before circulation, provide long-term multilineage reconstitution. Immunity 2001; 15:477-485.

30. Yoder MC, Hiatt K, Dutt P et al. Characterization of definitive lymphohematopoietic stem cells in the day 9 murine yolk sac. Immunity 1997; 7:335-344.

31. Yoder M, Cumming J, Hiatt K et al. A novel method of myeloablation to enhance engraftment of adult bone marrow cells in newborn mice. Biol Blood Marrow Transplant 1996; 2:59-67.

32. Yoder M, Hiatt K. Engraftment of embryonic hematopoietic cells in conditioned newborn recipients. Blood 1997; 89:2176-2183.

33. Yoder M, Hiatt K, Mukherjee P. In vivo repopulating hematopoietic stem cells are present in the murine yolk sac at day 9.0 postcoitus. Proc Natl Acad Sci USA 1997; 94:6776-6780.

34. Sato T, Laver J, Ogawa M. Reversible expression of CD34 by murine hematopoietic stem cells. Blood 1999; 94:2548-2554.
35. Ito T, Tajima F, Ogawa M. Developmental changes of CD34 expression by murine hematopoietic stem cells. Exp Hematol 2000; 28:1269-1273.
36. Zeigler F, Bennett B, Jordan C et al. Cellular and molecular characterization of the role of the flk-2/flt-3 receptor tyrosine kinase in hematopoietic stem cells. Blood 1994; 84:2422-2430.
37. Cumano A, Dieterlen-Lievre F, Godin I. Lymphoid potential, probed before circulation in mouse, is restricted to caudal intraembryonic splanchnopleura. Cell 1996; 86:907-916.
38. Matsuoka S, Tsuji K, Hisakawa H et al. Generation of definitive hematopoietic stem cells from murine early yolk sac and paraaortic splanchnopleura by aorta-gonad-mesonephros region-derived stromal cells. Blood 2001; 98:6-12.
39. Kyba M, Perlingeiro R, Daley GQ. HoxB4 confers definitive lymphoid-myeloid engraftment potential on embryonic stem cell and yolk sac hematopoietic progenitors. Cell 2002; 109:29-37.
40. Xu M, Tsuji K, Ueda T. Stimulation of mouse and human primitive hematopoiesis by murine embryonic aorta-gonad-mesonephros-derived stromal cell lines. Blood 1998; 92:2032-2040.
41. Godin I, Garcia-Porrero J, Coutinho A et al. Para-aortic splanchnopleura from early mouse embryos contains B1a cell progenitors. Nature 1993; 364:67-70.
42. Godin I, Dieterlen-Lievre F, Cumano A. Emergence of multipotent hemopoietic cells in the yolk sac and paraaortic splanchnopleura in mouse embryos, beginning at 8.5 days postcoitus. Proc Natl Acad Sci USA 1995; 92:773-777.
43. Cumano A, Godin I. Pluripotent hematopoietic stem cell development during embryogenesis. Curr Opin Immunol 2001; 13:166-171.
44. North T, Gu T, Stacy T et al. Cbfa2 is required for the formation of intra-aortic hematopoietic clusters. Development 1999; 126:2563-2575.
45. de Bruijn M, Speck N, Peeters M et al. Definitive hematopoietic stem cells first develop within the major arterial regions of the mouse embryo. EMBO J 2000; 19:2465-2474.
46. Godin I, Garcia-Porrero JA, Dieterlen-Lievre F et al. Stem cell emergence and hemopoietic activity are incompatible in mouse intraembryonic sites. J Exp Med 1999; 190:43-52.
47. North T, de Bruijn M, Stacy T et al. Runx1 expression marks long-term repopulating hematopoietic stem cells in the midgestation mouse embryo. Immunity 2002; 16:661-672.
48. Nishikawa S, Nishikawa S, Hirashima M et al. Progressive lineage analysis by cell sorting and culture identifies FLK 1+ VE-cadherin+ cells at a diverging point of endothelial and hematopoietic lineages. Development 1998; 125:1747-1757.
49. Nishikawa S-I, Nishikawa S, Kawamoto H et al. In vitro generation of lymphohematopoietic cells from endothelial cells purified from murine embryos. Immunity 1998; 8:761-769.
50. Hara T, Nakano Y-K, Tanaka M et al. Identification of podocalyxin-like protein 1 as a novel cell surface marker for hemangioblasts in the murine aorta-gonad-mesonephros region. Immunity 1999; 11:567-578.
51. de Bruijn M, Ma X, Robin C et al. Hematopoietic stem cells localize to the endothelial cell layer in the midgestation mouse aorta. Immunity 2002; 16:673-683.
52. Palis J, Robertson S, Kennedy M et al. Development of erythroid and myeloid progenitors in the yolk sac and embryo proper of the mouse. Development 1999; 126:5073-5081.
53. Palis J, Yoder M. Yolk Sac Hematopoiesis-The first blood cells of mouse and man. Exp Hematol 2001; 29:927-936.

CHAPTER 7

The Endothelium:

The Cradle of Definitive Hematopoiesis?

Katrin Ottersbach and Elaine Dzierzak

Abstract

It has been more than 80 years since a lineage relationship between hematopoietic and endothelial cells in the form of a bipotential precursor—the hemangioblast—was first proposed. Evidence has accumulated that supports the existence of such a cell, however, identification and isolation of hemangioblasts from embryos has so far not been achieved. The situation has been further complicated by the suggestion that different types of hemangioblasts with varying degrees of maturity exist, some of which may be restricted to either primitive or definitive hematopoietic lineages. Furthermore, recent work has pointed to a specialized group of endothelial cells that possess hemogenic potential and may represent an intermediate step in the generation of definitive hematopoietic stem cells. The following chapter provides a brief summary of recent developments in the investigation into the origins of hematopoiesis.

Background

It has long been established that the murine adult hematopoietic system derives from a small cohort of self-renewing hematopoietic stem cells (HSCs) that resides in the bone marrow. These stem cells provide the organism with a life-long supply of blood cells of all the different lineages, which are generated via a number of intermediate progenitors that become progressively more restricted in their self-renewal and differentiation potential. The adult hematopoietic system can thus be regarded as a hierarchy of cells with HSCs forming the base from which progenitors and mature cells are produced in a unidirectional, linear fashion within each lineage (reviewed in ref. 1).

Blood generation within the embryo, however, presents a more complicated picture. Hematopoiesis at early stages of ontogeny cannot be explained with the simple hierarchical model found in the adult. Instead, embryonic hematopoiesis is characterized by the emergence of several different progenitor populations, the origin and relationship of which are not well understood (reviewed in refs. 2, 3). Broadly speaking, hematopoiesis in the embryo can be defined by two consecutive waves of hematopoietic activity, a primitive and a definitive wave. The primitive wave of hematopoietic cells, which is transient, includes large, nucleated erythroblasts expressing embryonic globin genes and a population of primitive macrophages.[4-6] These populations are not only distinct from their definitive counterparts with respect to the sets of genes they express, but they also mature more rapidly in order to meet the metabolic demands of the developing embryo. Primitive erythroblasts are the first hematopoietic cells

---

Hematopoietic Stem Cell Development, edited by Isabelle Godin and Ana Cumano.
©2006 Eurekah.com and Kluwer Academic / Plenum Publishers.

found during development, and they are exclusively generated within the yolk sac at embryonic stage 7.5 (E7.5).

The first definitive, adult-type HSCs, as defined by their ability to replenish the adult hematopoietic system of an irradiated recipient, are detected within the embryo in a region that encompasses the developing aorta, gonads and mesonephros (AGM) starting at E10.5.[7,8] Between these two developmental time points, a number of other multipotent progenitors can be detected both within the yolk sac and the embryo proper (9-11 and reviewed in ref. 3). The precise function of these progenitors is as yet unknown and it is uncertain whether they persist into adult life or serve as precursors for the adult-type HSCs. Two independent studies have identified hematopoietic precursor cells that arise in the conceptus before E10.5 and that can, to some extent, provide long term reconstitution when transplanted into a more permissive environment. For example, CD34+c-kit+ cells isolated from the E9 yolk sac or P-Sp (paraaortic splanchnopleura – the developmental precursor of the AGM region) can repopulate neonatal mice when injected into the fetal liver at birth.[10] Similarly, cells from the E8.5 yolk sac and P-Sp can repopulate to low levels when transplanted into alymphoid RAG2γc$^{-/-}$ mice.[12] These studies imply that pre-HSCs exist in the embryo at earlier time points and that these cells lack certain properties that allow them to survive and proliferate in the adult environment. It remains to be shown what the precise nature and function of these cells is and whether they are indeed the immediate precursors of adult-type HSCs.

One day after adult-type HSCs are found in the AGM, they can also be detected in the yolk sac. Since circulation in the embryo commences at E8.5, it has long been debated whether the yolk sac has the capacity to generate and/or expand HSCs in situ or whether it is colonized. Recent evidence, based on explant cultures, suggests that in addition to the AGM, HSCs may indeed be generated and/or expanded in the yolk sac and that both AGM- and YS-derived HSCs may be required to provide a sufficient supply of these cells for the colonization of the fetal liver (FL), the principal hematopoietic organ until birth.[13]

The Emergence of Definitive HSCs: Origin and Regulation

Transplantation studies carried out in our group determined that the first murine HSCs localize to the major vessels of the developing mouse embryo, the dorsal aorta and the vitelline and umbilical vessels.[14] Immunohistochemical analyses in birds,[15] human,[16] mouse,[17] Xenopus[18] and zebrafish[19] have identified clusters of cells associated with the ventral endothelial wall of the dorsal aorta and the vitelline and umbilical arteries. These express a number of hematopoietic and endothelial markers, which include CD34,[16,20] CD31,[21] ckit,[22,23] CD45,[24] AA4.1,[25] endomucin,[26] CD164 in the human[27] and Flk1 (not in the avian embryo).[28] They also express mRNA for hematopoiesis-associated transcription factor genes such as SCL, c-myb, GATA2, GATA3, Lmo2 and AML1.[18,22,29-31] Some of the surface markers, namely CD31, CD34 and CD45, are also found on another group of hematopoietic cells which are localized in aggregates/foci in the dorsal mesentery ventral to the dorsal aorta in the mouse (17, 20 and reviewed in ref. 32) and chicken.[15,33] The cell clusters in both of these anatomical locations have been proposed to contain the first hematopoietic stem or progenitor cells. Delineating the events that lead to their production has therefore been the subject of intense investigation.

One model, based on chick embryo data, proposes that the ventral wall of the aorta contains endothelial cells with the capacity (for a brief period during development) to generate blood cells, which then bud off into the lumen of the aorta or migrate into the dorsal mesentery where they form the paraaortic foci. Others, using mouse models, have suggested that HSCs or pre-HSCs are generated within the mesenchyme just beneath the aorta and rapidly migrate through the endothelial layer into the aortic lumen.

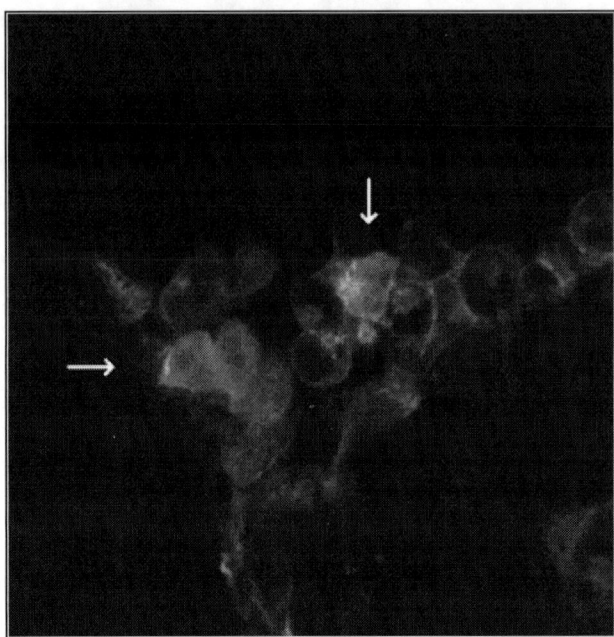

Figure 1. Ventral part of the dorsal aorta of a 39 SP Ly-6A-GFP embryo at the entry point of the vitelline artery. The section was stained with an antibody against CD34 (white cell surface staining or red in the color version). The white arrows mark GFP (cytosolic; bright green in the color version)-expressing cells (3 on the left, 1 at the top) which co-express CD34.

Thus, in order to further investigate the exact location of the first HSCs in the mouse embryo, we have placed the green fluorescent protein (GFP) under the transcriptional regulatory elements of the well-known HSC marker Sca-1 (Ly-6A). The transgenic animals were shown to express GFP in all long-term repopulating HSCs in the adult as well as in the embryo, as demonstrated by transplantation of sorted cells into irradiated adult recipients.[34,35] Sections of transgenic embryos at various stages revealed that GFP expression in the dorsal aorta initiates at E9 (24 somite pairs) and, at this stage, is found exclusively in endothelial cells in the ventral portion of the dorsal aorta. As development proceeds, the number of GFP-expressing endothelial cells increases until they are seen around the entire circumference of the aorta by late E11. At that point, GFP+ cells are also observed within clusters of cells attached to the ventral endothelium and as a few round circulating cells. Interestingly, larger clusters, which are especially prominent at the junction between the dorsal aorta and the vitelline artery, are never entirely composed of GFP-expressing cells (Fig. 1). Heterogeneity within the clusters has been demonstrated previously in studies using antibodies against von Willebrand factor[20] or CD41[36] and may reflect some degree of maturation occurring within these groups of cells. Staining with other markers for hematopoietic and endothelial cells revealed that GFP+ cells in the endothelium and clusters co-express CD31 and CD34 (Fig. 1). We did not, however, detect any aggregates of GFP+ cells in the subaortic mesenchyme.

GFP+ cells were also found in the endothelial layer of the vitelline and umbilical vessels, following the same temporal pattern as in the dorsal aorta. However, the expression of GFP in the YS was not observed until almost a day later and was then restricted to the larger vessels. A summary of these expression data is presented in Table 1. Taken together, our results obtained

Table 1. *Summary of the expression pattern of the Ly-6A-GFP transgene in hematopoietic tissues during specific developmental stages*

	Aorta	Umbilical/Vitelline Vessels	Yolk Sac
Day 9 (>24SP)	+	++	-
Day 10	++	++	+
Day 11	+++	+++	++

Data was obtained by inspection of whole embryos or transversal sections through fixed embryos. The number of GFP-expressing cells is represented by – (none), + (lowest) to +++ (highest).

with the Ly6A-GFP transgenic mice suggest that HSCs and/or their precursors are generated within the endothelial layer of the dorsal aorta. Since all HSCs express GFP in our system, but not all GFP+ cells are in fact HSCs, we cannot say at this stage whether the repopulating cells are located in the clusters or in the endothelial layer. However, recent data by North et al[37] point to the presence of HSCs in both the hematopoietic and, to a lesser extent, in the endothelial cell populations. Transplantation of GFP+ cells from early E10 embryos (31-34 somite pairs) resulted in limited repopulation (1-2%), hinting at the possibility that the transgene also marks multipotential progenitors and/or pre-HSCs. Our results again highlight the fact that a close developmental relationship exists between endothelial and hematopoietic cells, as both of these cell types are marked by the transgene. In order to put our results into context, a brief review of this close relationship is given below.

Hemangioblast or Hemogenic Endothelial Cells?

The existence of a common bipotential precursor for endothelial and hematopoietic cells was originally suggested in 1920,[38] a cell that is now referred to as the hemangioblast.[39] Since then further supporting evidence for a lineage relationship has accumulated and includes the observation that the two cell types share a number of surface markers, such as CD31,[40] CD34,[41] Flk1,[42,43] Flt1,[44] VE-Cadherin[37] and Tie2,[45] and that null mutations in a number of genes, such as Flk1,[46] SCL[47-49] and Tie2,[45] affect both cell lineages. Furthermore, endothelial and hematopoietic cells are the first differentiated lineages that appear during development. They emerge simultaneously in close proximity to each other from a seemingly homogeneous population of mesodermal precursors in the yolk sac blood islands of the E7.5 murine conceptus.

In vitro studies using the embryonic stem (ES) cell system have provided further evidence for the existence of the hemangioblast. Blast colony forming cells (BL-CFCs), derived from ES cells and grown in the presence of the Flk1 ligand VEGF, can give rise to both hematopoietic and endothelial cells in culture.[50] Flk1 serves as a marker for the hemangioblast, since sorted Flk1+ ES cells can give rise to hematopoietic and endothelial cells in vitro.[51,52] A functional role for Flk1 in hemangioblast emergence had already been suggested based on the phenotype of Flk1-deficient mice, which have a severe defect in the endothelial as well as the hematopoietic lineage.[28,46] However, it is not clear how Flk1 is involved in hemangioblast specification, since hematopoietic cells can be derived from Flk1$^{-/-}$ ES cells. Also, Flk1$^{-/-}$ ES cells express endothelial-specific genes.[53,54] The fact that Flk1$^{-/-}$ cells that express lacZ under the Flk1 regulatory elements are found in aberrant positions within the embryo,[28,46,54] together with recent findings in Drosophila,[55] strongly suggests that Flk1 may guide the migration of mesodermal progenitors along a VEGF gradient to the appropriate location where endothelial and hematopoietic maturation can occur. Since Flk1 continues to be expressed in endothelial cells, it appears that it also has an instructive role in endothelial differentiation, whereas it plays only a permissive role in hematopoietic differentiation.[53]

The transcription factor SCL is likely to be another major regulator of hemangioblast generation since its absence results in major defects in vasculogenesis and a complete block in the entire hematopoietic program.[47-49] It may act at a slightly later stage than Flk1, as ES cells differentiated in vitro express Flk1 before they start expressing SCL.[51] Nonetheless, activation of both of these genes is required for the differentiation of ES cells into endothelial and hematopoietic cells.

Despite all these studies, it has been thus far impossible to prove the existence in vivo of such a mesodermal bipotential precursor on a single cell level.

The studies described above still leave open the question concerning the initiation of definitive hematopoiesis. It was long believed that the yolk sac was the only site where hematopoietic cells were generated and that it was those cells that colonized the fetal liver.[56] However, studies on avian embryos[57,58] and subsequently murine embryos firmly established that an independent source within the embryo also had the capacity to autonomously generate hematopoietic cells.[8,9] As mentioned above, intraembryonic hematopoietic activity was detected at the site of the developing major arteries beginning at E8.5,[9] with adult-type HSCs being produced starting at E10.5.[8,14] The fact that endothelial cells are already present at those time points makes it difficult to link the emergence of intra-embryonic blood cells to a population of hemangioblasts. It may well be possible that the intraembryonic microenvironment does not initially permit the production of blood cells, thus forcing hemangioblast differentiation exclusively along the endothelial pathway. Indeed, it was demonstrated that Flk1+ ES cells that can give rise to hematopoietic as well as endothelial cells in vitro would only produce the latter when VEGF was added to the cultures.[42] Similarly, AGM cells cultured in the presence of VEGF produced fewer hematopoietic cells when subsequently plated in methylcellulose.[59] Local secretion of such an inhibitory signal may thus delay the onset of hematopoiesis in the P-Sp region.

Recent studies have proposed that definitive hematopoietic cells are directly derived from an endothelial precursor (reviewed in refs. 60-62). This 'hemogenic' endothelium is thought to comprise a specialized subpopulation of endothelial cells that can give rise to hematopoietic cells in a spatially and temporally restricted fashion. These cells are thought to be located on the ventral side of the dorsal aorta, as it is evident from embryo sections that the ventral endothelium has a morphology that is distinct from the endothelium of the dorsal side and from that of other blood vessels.[20,24,30,33,63] The normal layer of flat endothelial cells is disrupted in the ventral location and replaced by a thicker and less organized population of rounded cells. Clusters of hematopoietic cells seem to be incorporated in the endothelium, with individual cells appearing to bud off into the lumen. Cells from the clusters may also migrate into the underlying mesenchyme. The change from normal endothelium to hemogenic endothelium in the avian embryo is also associated with a change in marker expression, as endothelial cells in the cluster-bearing areas lose Flk1 expression and start expressing CD45 when they mature into hematopoietic cells.[33] Similarly, Flk1 and c-kit expression is mutually exclusive in hemogenic areas of the dorsal aorta in murine embryos.[23]

Lineage studies have been employed to determine whether endothelial cells can indeed give rise to hematopoietic cells (see chapter by Jaffredo). Prior to the appearance of hematopoietic cells at E2, endothelial cells in the avian embryo were labeled either with dye-coupled LDL molecules or with a lacZ-expressing retrovirus.[33,64] Initially, the label was only detected in endothelial cells. However, one day later, labeled CD45+ hematopoietic cells were observed within the intra-aortic clusters, and by E6, labeled CD45+ cells were also found in the para-aortic foci. These data provide evidence for the developmental relationship between endothelial and hematopoietic cells with the former giving rise to the latter and thus passing on the marker. However, the timing of the retroviral infection is crucial. Infection of endothelial cells at E4 did not result in labeling of para-aortic foci implying that these specialized endothelial cells only transiently possess hemogenic potential. This is supported by the observation that the CD45$^+$ portions of the ventral endothelium disappear as it reverts back to a uniform Flk1$^+$ phenotype by E4.[33] It is of note that groups of flat Flk1$^+$endothelial cells were never found intermixed with round, budding hematopoietic cells. This, together with data from Eichmann et al[42] which showed that Flk1$^+$ cells from the chicken blastodisc could give rise to endothelial or hematopoietic cells, but never to both in the same culture, has so far precluded the identification of a bipotential hemangioblast in avian embryos.

The ability of endothelial cells to give rise to blood cells has also been demonstrated in vitro with VE-Cadherin$^+$ cells sorted from the murine yolk sac and embryo proper and cultured on OP9 stromal cells in the presence of interleukin-7.[65] These cells were identified as endothelial cells by CD31 expression, LDL uptake and lack of expression of the hematopoietic markers CD45 and Ter119. When isolated from the yolk sac and embryo proper at E8.5, these cells were restricted to an erythroid-myeloid potential, while at E9.5 and E10, they were also able to generate B cells and T cells, respectively. These hemogenic endothelial cells had the same hematopoietic potential regardless of whether they originated from the yolk sac or embryo proper. Hematopoietic cells (a mixture of sorted CD45$^+$ and Ter119$^+$ cells) from the embryo proper at different time points were found to have the same potential as endothelial cells from the same region. However, yolk sac-derived hematopoietic cells acquired their lymphoid potential at later developmental stages than the yolk sac endothelial cells. Thus, it was suggested that mesoderm to endothelial cell differentiation (including hemogenic endothelial cells) occurs in a similar fashion in the yolk sac and the embryo proper, while differentiation from endothelial to hematopoietic cells is regulated differently in those two compartments.[65] VE-cadherin$^+$ CD45$^-$ hemogenic endothelial cells from E9.5 and E10.5 embryos also possess partial reconstitution activity when injected into the livers of conditioned newborn SCID mice.[66]

VE-cadherin$^+$ cells were also detected during the in vitro hematopoietic differentiation of ES cells.[52] These may be a more mature subpopulation of hemangioblastic Flk1$^+$ ES cells at the diverging point of the endothelial and hematopoietic lineages. Further studies suggest that the expression of α4 integrin marks the step at which commitment to the hematopoietic lineage occurs.[67]

The Tie2$^+$ fraction in E10.5 AGM cells also harbors hemogenic endothelial cells,[45,68] as shown by the production of hematopoietic as well as endothelial cells when cultured on OP9 stromal cells. After 14 days of culture, round cells were found growing on an endothelial network and were confirmed to be hematopoietic by staining for specific markers and by methylcellulose assays.[68] No such activity was found in the Tie2$^-$ fraction.

Our own studies using the Ly-6A-GFP transgenic mice lend further support to the notion that definitive hematopoiesis originates from the endothelial compartment. Expression of the transgene initiates in endothelial cells at the ventral aspect of the dorsal aorta at day 9 and remains restricted to endothelial cells until the emergence of clusters, which also harbor GFP$^+$ cells. Thus, the transgene may mark endothelial cells that have acquired hematopoietic potential and that are the direct precursors of HSCs. It would be of great interest to test whether these early GFP-expressing endothelial cells have neonatal repopulating activity.

Hemogenic endothelial cells have also been identified in the human embryo and shown to display a CD34$^+$ CD45$^-$ or CD31$^+$ CD45$^-$ phenotype.[21] Like their murine counterparts, they require a culture step on stromal cells (MS-5 in this case) in order to be able to differentiate into hematopoietic cells. They were detected in the yolk sac, embryonic AGM, embryonic and fetal liver and in the fetal bone marrow, but not in the fetal aorta or other non-hemogenic sites. Unlike their intraembryonic counterparts, hemogenic endothelial cells sorted from the human yolk sac are not endowed with a lymphoid potential.

It is still unknown how these ventral endothelial cells acquire their hemogenic potential. Grafting studies in chick/quail chimaeras demonstrate that the ventral endothelium and the dorsolateral endothelium of the dorsal aorta have distinct mesodermal origins. The ventral endothelium is derived from the splanchnopleural mesoderm whereas the dorsal endothelium originates from the somatopleural mesoderm.[69,70] Since only the splanchnopleural mesoderm has been shown to possess endothelial as well as hematopoietic potential, the hematopoietic potential of the ventral mesoderm may be cell-autonomous. However, additional local signals also seem to be required, since splanchnopleural mesoderm-derived cells will only give rise to blood cells in the appropriate location.[69] Such a signal may be provided by the synergistic action of the transcription factors GATA3 and Lmo2 which are co-expressed in the underlying mesenchyme prior to the emergence of hematopoietic precursors.[29] Other reports have also highlighted local changes in the mesoderm underneath the clusters such as a denser appearance,[33,71] an accumulation of smooth muscle actin-positive cells[24,72] and a high expression of tenascin-C.[71] It still remains to be shown whether any of these local variations in the mesoderm have an influence on the generation of hematopoietic cells.

There is accumulating evidence for a role of the transcription factor AML1 in the transition of hemogenic endothelial cells to hematopoietic stem and/or progenitor cells. Mice that are deficient for AML1 have a complete block in definitive hematopoiesis, while primitive yolk sac hematopoiesis is unaffected.[73,74] AML1 expression in the E10.5 AGM region (as followed by the expression of a lacZ reporter in a mutant allele) is in hematopoietic cells in the lumen of the dorsal aorta, as well as in cells attached to the endothelium.[30] Endothelial cells in the dorsal aorta are positive, the majority of which are located on the ventral side of the aorta. Interestingly, the expression of AML1 is transient in these cells and coincides with HSC emergence. A few lacZ+ cells are also detected in the underlying mesenchyme and in the mesonephros. In AML1-/- mice, the same AML1-lacZ expression pattern in endothelial and mesenchymal cells was found at E9.5. However, by E10.5, the number of lacZ-expressing endothelial cells is markedly decreased, with all of them displaying a flat, elongated morphology. In addition, there is an accumulation of positive cells in the mesenchyme underneath. No hematopoietic clusters are found in AML1-/- embryos.[30] AML1 may thus specify a subset of endothelial cells to form clusters. A similar role for AML1 at the hemogenic endothelial cell stage was suggested by another group who found AML1 transcripts in sorted VE-Cadherin+ CD45- Ter119- cells of E10.5 embryos.[75] Furthermore, VE-Cadherin+ CD45- Ter119- sorted cells from AML1-/- E10.5 yolk sacs and AGM regions are unable to produce hematopoietic cells when cultured on OP9 stromal cells, unlike their wild-type counterparts. In contrast, endothelial cells appear normal, although their numbers are slightly increased. More recent data places AML1 expression at an even earlier stage.[37] Through a combination of markers that allowed discrimination of mesenchymal cells, endothelial cells and hematopoietic clusters, together with transplantation assays for the detection of HSCs, it was shown that HSCs in AML1+/- embryos are located in the mesenchyme, the endothelium as well as in the clusters, whereas in wild-type embryos, they are almost exclusively within the clusters. It thus appears that a lower dose of AML1 causes a delay in the transition of mesenchymal precursors into hemogenic endothelial cells and further into clusters. As a result, HSCs accumulate at intermediate stages. It cannot be entirely ruled out, however, that a lower dose of AML1 affects the expression of some of the markers that define the different cell populations.

Concluding Remarks

The studies described above, together with our own data from the Ly-6A-GFP transgenic mice, provide compelling evidence for HSCs being derived from a population of hemogenic endothelial cells. Should this then be regarded as the hemangioblast that gives rise to the definitive hematopoietic lineage? The generally accepted definition for the hemangioblast is a bipotential mesodermal progenitor for the endothelial and hematopoietic lineage. It is becoming increasingly apparent that hemogenic endothelial cells may represent an intermediate step between a hemangioblast-like cell, which may be located in the mesenchyme underneath the aorta, and HSCs. In our transgenic mouse model, these mesodermal progenitors are likely to be GFP⁻ and to switch on the expression of the transgene when they integrate in and/or migrate through the ventral endothelium of the aorta where they appear as hemogenic endothelial cells. ES cell studies suggest that there are most likely different populations of hemangioblasts, a primitive one that gives rise to endothelial cells and primitive blood cells in the early yolk sac, a definitive one that produces endothelial and definitive hematopoietic cells in the embryo proper (and possibly also in the yolk sac at later stages), and a third one that is the precursor of the first two. The common precursor for primitive and definitive hematopoiesis identified by Kennedy et al[76] may well represent such a pre-hemangioblast. Recent work on Xenopus embryos has identified two different progenitor populations for adult and embryonic blood that also produce endothelial cells (see chapter by Ciau-Uitz).[77] These arise in different parts of the embryo, have different migratory pathways and respond differently to BMP4 and thus may represent the two more mature hemangioblast populations. The definitive hemangioblast subset, at least in the ES cell system, may be distinguished by the expression of AML1,[78] VE-Cadherin,[79] and/or PCLP1.[80] The possibility that there are two distinct hemangioblast populations for definitive and primitive hematopoiesis may also explain why the two hematopoietic lineages require different genetic programs, which nevertheless seem to be activated simultaneously.[81]

This model of different hemangioblast populations giving rise to primitive and definitive hematopoiesis still leaves a lot of questions unanswered. Why have these two populations evolved and why does definitive hematopoiesis have to go through an endothelial intermediate? It has been suggested that the endothelium provides a suitable environment in which multipotential hematopoietic progenitors can be generated.[62] Another important question is how long these hemangioblasts and hemogenic endothelial cells persist. Although they appear to be transient in the dorsal aorta,[64] they may well go on to colonize the fetal liver and even the bone marrow.[21,82] The answers to these questions will require detailed lineage tracing and fate mapping studies.

Acknowledgements

We would like to thank all the members of the lab for their helpful discussion and the following funding agencies for their general support, The Wellcome Trust (GR063331MA) and the National Institutes of Health (NIH RO1 DK51077).

References

1. Graham GJ, Wright EG. Hematopoietic stem cells: their heterogeneity and regulation. Int J Exp Pathol 1997; 78(4):197-218.
2. Dzierzak E, Medvinsky A, de Bruijn M. Qualitative and quantitative aspects of haematopoietic cell development in the mammalian embryo. Immunol Today 1998; 19(5):228-36.
3. Keller G, Lacaud G, Robertson S. Development of the hematopoietic system in the mouse. Exp Hematol 1999; 27(5):777-87.
4. Brotherton TW, Chui DH, Gauldie J et al. Hemoglobin ontogeny during normal mouse fetal development. Proc Natl Acad Sci USA 1979; 76(6):2853-7.

5. Cline MJ, Moore MA. Embryonic origin of the mouse macrophage. Blood 1972; 39(6):842-9.

6. Faust N, Huber MC, Sippel AE et al. Different macrophage populations develop from embryonic/ fetal and adult hematopoietic tissues. Exp Hematol 1997; 25(5):432-44.

7. Muller AM, Medvinsky A, Strouboulis J et al. Development of hematopoietic stem cell activity in the mouse embryo. Immunity 1994; 1(4):291-301.

8. Medvinsky A, Dzierzak E. Definitive hematopoiesis is autonomously initiated by the AGM region. Cell 1996; 86(6):897-906.

9. Cumano A, Dieterlen-Lievre F, Godin I. Lymphoid potential, probed before circulation in mouse, is restricted to caudal intraembryonic splanchnopleura. Cell 1996; 86(6):907-16.

10. Yoder MC, Hiatt K, Dutt P et al. Characterization of definitive lymphohematopoietic stem cells in the day 9 murine yolk sac. Immunity 1997; 7(3):335-44.

11. Palis J, Robertson S, Kennedy M et al. Development of erythroid and myeloid progenitors in the yolk sac and embryo proper of the mouse. Development 1999; 126(22):5073-84.

12. Cumano A, Ferraz JC, Klaine M et al. Intraembryonic, but not yolk sac hematopoietic precursors, isolated before circulation, provide long-term multilineage reconstitution. Immunity 2001; 15(3):477-85.

13. Kumaravelu P, Hook L, Morrison AM et al. Quantitative developmental anatomy of definitive haematopoietic stem cells/long-term repopulating units (HSC/RUs): role of the aorta-gonad- mesonephros (AGM) region and the yolk sac in colonisation of the mouse embryonic liver. Development 2002; 129(21):4891-9.

14. de Bruijn MF, Speck NA, Peeters MC et al. Definitive hematopoietic stem cells first develop within the major arterial regions of the mouse embryo. EMBO J 2000; 19(11):2465-74.

15. Dieterlen-Lievre F, Martin C. Diffuse intraembryonic hemopoiesis in normal and chimeric avian development. Dev Biol 1981;88(1):180-91.

16. Tavian M, Coulombel L, Luton D et al. Aorta-associated CD34+ hematopoietic cells in the early human embryo. Blood 1996; 87(1):67-72.

17. Garcia-Porrero JA, Godin IE, Dieterlen-Lievre F. Potential intraembryonic hemogenic sites at pre-liver stages in the mouse. Anat Embryol (Berl) 1995; 192(5):425-35.

18. Ciau-Uitz A, Walmsley M, Patient R. Distinct origins of adult and embryonic blood in Xenopus. Cell 2000; 102(6):787-96.

19. Thompson MA, Ransom DG, Pratt SJ et al. The cloche and spadetail genes differentially affect hematopoiesis and vasculogenesis. Dev Biol 1998;197(2):248-69.

20. Garcia-Porrero JA, Manaia A, Jimeno J et al. Antigenic profiles of endothelial and hemopoietic lineages in murine intracmbryonic hemogenic sites. Dev Comp Immunol 1998; 22(3):303-19.

21. Oberlin E, Tavian M, Blazsek I et al. Blood-forming potential of vascular endothelium in the human embryo. Development 2002; 129(17):4147-57.

22. Labastie MC, Cortes F, Romeo PH et al. Molecular identity of hematopoietic precursor cells emerging in the human embryo. Blood 1998; 92(10):3624-35.

23. Yoshida H, Takakura N, Hirashima M et al. Hematopoietic tissues, as a playground of receptor tyrosine kinases of the PDGF-receptor family. Dev Comp Immunol 1998; 22(3):321-32.

24. Tavian M, Cortes F, Charbord P et al. Emergence of the haematopoietic system in the human embryo and foetus. Haematologica 1999; 84 Suppl EHA-4:1-3.

25. Petrenko O, Beavis A, Klaine M et al. The molecular characterization of the fetal stem cell marker AA4. Immunity 1999; 10(6):691-700.

26. Brachtendorf G, Kuhn A, Samulowitz U et al. Early expression of endomucin on endothelium of the mouse embryo and on putative hematopoietic clusters in the dorsal aorta. Dev Dyn 2001; 222(3):410-9.

27. Watt SM, Butler LH, Tavian M et al. Functionally defined CD164 epitopes are expressed on CD34(+) cells throughout ontogeny but display distinct distribution patterns in adult hematopoietic and nonhematopoietic tissues. Blood 2000; 95(10):3113-24.

28. Shalaby F, Ho J, Stanford WL et al. A requirement for Flk1 in primitive and definitive hematopoiesis and vasculogenesis. Cell 1997; 89(6):981-90.

29. Manaia A, Lemarchandel V, Klaine M et al. Lmo2 and GATA-3 associated expression in intraembryonic hemogenic sites. Development 2000; 127(3):643-53.

30. North T, Gu TL, Stacy T et al. Cbfa2 is required for the formation of intra-aortic hematopoietic clusters. Development 1999; 126(11):2563-75.

31. Vandenbunder B, Pardanaud L, Jaffredo T et al. Complementary patterns of expression of c-ets 1, c-myb and c-myc in the blood-forming system of the chick embryo. Development 1989; 107(2):265-74.

32. Godin I, Cumano A. The hare and the tortoise: an embryonic haematopoietic race. Nat Rev Immunol 2002; 2(8):593-604.

33. Jaffredo T, Gautier R, Eichmann A et al. Intraaortic hemopoietic cells are derived from endothelial cells during ontogeny. Development 1998; 125(22):4575-83.

34. de Bruijn MF, Ma X, Robin C et al. Hematopoietic stem cells localize to the endothelial cell layer in the midgestation mouse aorta. Immunity 2002; 16(5):673-83.

35. Ma X, Robin C, Ottersbach K et al. The Ly-6A (Sca-1) GFP Transgene is Expressed in all Adult Mouse Hematopoietic Stem Cells. Stem Cells 2002; 20(6):514-21.

36. Ody C, Vaigot P, Quere P et al. Glycoprotein IIb-IIIa is expressed on avian multilineage hematopoietic progenitor cells. Blood 1999; 93(9):2898-906.

37. North TE, de Bruijn MF, Stacy T et al. Runx1 expression marks long-term repopulating hematopoietic stem cells in the midgestation mouse embryo. Immunity 2002; 16(5):661-72.

38. Sabin F. Studies on the origin of blood vessels and of red blood corpuscules as seen in the living blastoderm of chicks during the second day of incubation. Contrib. Embryol. Carnegie Inst. of Washington 1920; 9:214-262.

39. Murray PDF. The Development in vitro of the Blood of the early Chick Embryo. Proc R Soc London 1932; 111:497-521.

40. Watt SM, Gschmeissner SE, Bates PA. PECAM-1: its expression and function as a cell adhesion molecule on hemopoietic and endothelial cells. Leuk Lymphoma 1995; 17(3-4):229-44.

41. Young PE, Baumhueter S, Lasky LA. The sialomucin CD34 is expressed on hematopoietic cells and blood vessels during murine development. Blood 1995; 85(1):96-105.

42. Eichmann A, Corbel C, Nataf V et al. Ligand-dependent development of the endothelial and hemopoietic lineages from embryonic mesodermal cells expressing vascular endothelial growth factor receptor 2. Proc Natl Acad Sci USA 1997; 94(10):5141-6.

43. Kabrun N, Buhring HJ, Choi K et al. Flk-1 expression defines a population of early embryonic hematopoietic precursors. Development 1997; 124(10):2039-48.

44. Fong GH, Zhang L, Bryce DM et al. Increased hemangioblast commitment, not vascular disorganization, is the primary defect in flt-1 knock-out mice. Development 1999; 126(13):3015-25.

45. Takakura N, Huang XL, Naruse T et al. Critical role of the TIE2 endothelial cell receptor in the development of definitive hematopoiesis. Immunity 1998; 9(5):677-86.

46. Shalaby F, Rossant J, Yamaguchi TP et al. Failure of blood-island formation and vasculogenesis in Flk-1-deficient mice. Nature 1995; 376(6535):62-6.

47. Visvader JE, Fujiwara Y, Orkin SH. Unsuspected role for the T-cell leukemia protein SCL/tal-1 in vascular development. Genes Dev 1998; 12(4):473-9.

48. Shivdasani RA, Mayer EL, Orkin SH. Absence of blood formation in mice lacking the T-cell leukaemia oncoprotein tal-1/SCL. Nature 1995; 373(6513):432-4.

49. Robb L, Lyons I, Li R et al. Absence of yolk sac hematopoiesis from mice with a targeted disruption of the scl gene. Proc Natl Acad Sci USA 1995; 92(15):7075-9.

50. Choi K, Kennedy M, Kazarov A et al. A common precursor for hematopoietic and endothelial cells. Development 1998; 125(4):725-32.

51. Chung YS, Zhang WJ, Arentson E et al. Lineage analysis of the hemangioblast as defined by FLK1 and SCL expression. Development 2002; 129(23):5511-20.

52. Nishikawa SI, Nishikawa S, Hirashima M et al. Progressive lineage analysis by cell sorting and culture identifies FLK1+VE-cadherin+ cells at a diverging point of endothelial and hemopoietic lineages. Development 1998; 125(9):1747-57.

53. Hidaka M, Stanford WL, Bernstein A. Conditional requirement for the Flk-1 receptor in the in vitro generation of early hematopoietic cells. Proc Natl Acad Sci USA 1999; 96(13):7370-5.

54. Schuh AC, Faloon P, Hu QL et al. In vitro hematopoietic and endothelial potential of flk-1(-/-) embryonic stem cells and embryos. Proc Natl Acad Sci USA 1999; 96(5):2159-64.

55. Cho NK, Keyes L, Johnson E et al. Developmental control of blood cell migration by the Drosophila VEGF pathway. Cell 2002; 108(6):865-76.
56. Moore MA, Metcalf D. Ontogeny of the hematopoietic system: yolk sac origin of in vivo and in vitro colony forming cells in the developing mouse embryo. Br J Haematol 1970; 18(3):279-96.
57. Dieterlen-Lievre F. On the origin of hematopoietic stem cells in the avian embryo: an experimental approach. J Embryol Exp Morph 1975;33:607-619.
58. Lassila O, Martin C, Toivanen P et al. Erythropoiesis and lymphopoiesis in the chick yolk-sac-embryo chimeras: contribution of yolk sac and intraembryonic stem cells. Blood 1982; 59(2):377-81.
59. Minehata K, Mukouyama YS, Sekiguchi T et al. Macrophage colony stimulating factor modulates the development of hematopoiesis by stimulating the differentiation of endothelial cells in the AGM region. Blood 2002; 99(7):2360-8.
60. Dieterlen-Lievre F. Hematopoiesis: progenitors and their genetic program. Curr Biol 1998; 8(20):R727-30.
61. Nishikawa SI. A complex linkage in the developmental pathway of endothelial and hematopoietic cells. Curr Opin Cell Biol 2001; 13(6):673-8.
62. Ogawa M, Fraser S, Fujimoto T et al. Origin of hematopoietic progenitors during embryogenesis. Int Rev Immunol 2001; 20(1):21-44.
63. Wood HB, May G, Healy L et al. CD34 expression patterns during early mouse development are related to modes of blood vessel formation and reveal additional sites of hematopoiesis. Blood 1997; 90(6):2300-11.
64. Jaffredo T, Gautier R, Brajeul V et al. Tracing the progeny of the aortic hemangioblast in the avian embryo. Dev Biol 2000; 224(2):204-14.
65. Nishikawa SI, Nishikawa S, Kawamoto H et al. In vitro generation of lymphohematopoietic cells from endothelial cells purified from murine embryos. Immunity 1998; 8(6):761-9.
66. Fraser ST, Ogawa M, Yu RT et al. Definitive hematopoietic commitment within the embryonic vascular endothelial-cadherin(+) population. Exp Hematol 2002; 30(9):1070-8.
67. Ogawa M, Kizumoto M, Nishikawa S et al. Expression of alpha4-integrin defines the earliest precursor of hematopoietic cell lineage diverged from endothelial cells. Blood 1999; 93(4):1168-77.
68. Hamaguchi I, Huang XL, Takakura N et al. In vitro hematopoietic and endothelial cell development from cells expressing TEK receptor in murine aorta-gonad-mesonephros region. Blood 1999; 93(5):1549-56.
69. Pardanaud L, Luton D, Prigent M et al. Two distinct endothelial lineages in ontogeny, one of them related to hemopoiesis. Development 1996; 122(5):1363-71.
70. Pardanaud L, Dieterlen-Lievre F. Manipulation of the angiopoietic/hemangiopoietic commitment in the avian embryo. Development 1999; 126(4):617-27.
71. Marshall CJ, Moore RL, Thorogood P et al. Detailed characterization of the human aorta-gonad-mesonephros region reveals morphological polarity resembling a hematopoietic stromal layer. Dev Dyn 1999; 215(2):139-47.
72. Takahashi Y, Imanaka T, Takano T. Spatial and temporal pattern of smooth muscle cell differentiation during development of the vascular system in the mouse embryo. Anat Embryol (Berl) 1996; 194(5):515-26.
73. Okuda T, van Deursen J, Hiebert SW et al. AML1, the target of multiple chromosomal translocations in human leukemia, is essential for normal fetal liver hematopoiesis. Cell 1996; 84(2):321-30.
74. Wang Q, Stacy T, Binder M et al. Disruption of the Cbfa2 gene causes necrosis and hemorrhaging in the central nervous system and blocks definitive hematopoiesis. Proc Natl Acad Sci USA 1996; 93(8):3444-9.
75. Yokomizo T, Ogawa M, Osato M et al. Requirement of Runx1/AML1/PEBP2alphaB for the generation of haematopoietic cells from endothelial cells. Genes Cells 2001; 6(1):13-23.
76. Kennedy M, Firpo M, Choi K et al. A common precursor for primitive erythropoiesis and definitive haematopoiesis. Nature 1997; 386(6624):488-93.
77. Walmsley M, Ciau-Uitz A, Patient R. Adult and embryonic blood and endothelium derive from distinct precursor populations which are differentially programmed by BMP in Xenopus. Development 2002; 129(24):5683-95.

78. Lacaud G, Gore L, Kennedy M et al. Runx1 is essential for hematopoietic commitment at the hemangioblast stage of development in vitro. Blood 2002; 100(2):458-66.
79. Fujimoto T, Ogawa M, Minegishi N et al. Step-wise divergence of primitive and definitive haematopoietic and endothelial cell lineages during embryonic stem cell differentiation. Genes Cells 2001; 6(12):1113-27.
80. Hara T, Nakano Y, Tanaka M et al. Identification of podocalyxin-like protein 1 as a novel cell surface marker for hemangioblasts in the murine aorta-gonad-mesonephros region. Immunity 1999; 11(5):567-78.
81. Endoh M, Ogawa M, Orkin S et al. SCL/tal-1-dependent process determines a competence to select the definitive hematopoietic lineage prior to endothelial differentiation. EMBO J 2002; 21(24):6700-8.
82. Gunsilius E, Duba HC, Petzer AL et al. Evidence from a leukaemia model for maintenance of vascular endothelium by bone-marrow-derived endothelial cells. Lancet 2000; 355(9216):1688-91.

CHAPTER 8

Endothelial and Hematopoietic Cells in the Intraembryonic Compartment

Fumio Arai and Toshio Suda

Abstract

The development of endothelial cells and hematopoietic cells is closely related at embryonic stages, and both cells share a common precursor, the hemangioblast. Differentiation into hematopoietic and endothelial lineages begins with proliferation of a single layer of mesodermal cells in the yolk sac resulting in formation of cell clusters, the blood islands. Cells at the periphery of these mesodermal aggregations differentiate into angioblasts. Blood vessels are formed by two different steps, vasculogenesis and angiogenesis. Cells in the interior of blood islands become primitive hematopoietic cells (HCs). Definitive hematopoiesis develops from the P-Sp region as early as E8.5. By E10.5, hematopoietic stem cells (HSCs) originate in the aorta-gonad-mesonephros (AGM) region. HSCs from this region colonize the fetal liver and then move to the spleen and bone marrow. In the embryo, HCs form clusters closely associated with and often adhering to endothelial cells on the ventral surface (floor) of the aorta. HCs that form these clusters also appear to be derived from the P-Sp. In addition, hematopoietic clusters have also been identified in the vitelline and umbilical arteries, indicating that intraembryonic hematopoietic development is associated with the major arterial region of the embryo.

Hematopoietic and endothelial lineages express several genes in common, such as VEGFR2, Tie2, and SCL/Tal-1. The fact that both lineages co-express common genes, many of which encode growth factor receptors or transcription factors, is not only consistent with the notion that they share a common precursor, but also suggests that similar molecular programs and growth regulatory mechanisms are involved in their development. Moreover, it has been reported that HSCs produce Ang1, which promotes angiogenesis.

Development of Vasculature

Blood vessels in the embryo form through two different processes; vasculogenesis and angiogenesis (Fig. 1). The earliest stages of vascular development, vasculogenesis, include differentiation, expansion, and coalescence of vascular endothelial cell precursors (EPCs) (angioblasts) into the initial vascular network.[1] This initial network consists solely of vascular endothelial cells (ECs) that have formed similar sized interconnected vessels, and are thus referred to as the primary capillary plexus. The primary plexus is then remodeled by a process

---

Hematopoietic Stem Cell Development, edited by Isabelle Godin and Ana Cumano.
©2006 Eurekah.com and Kluwer Academic / Plenum Publishers.

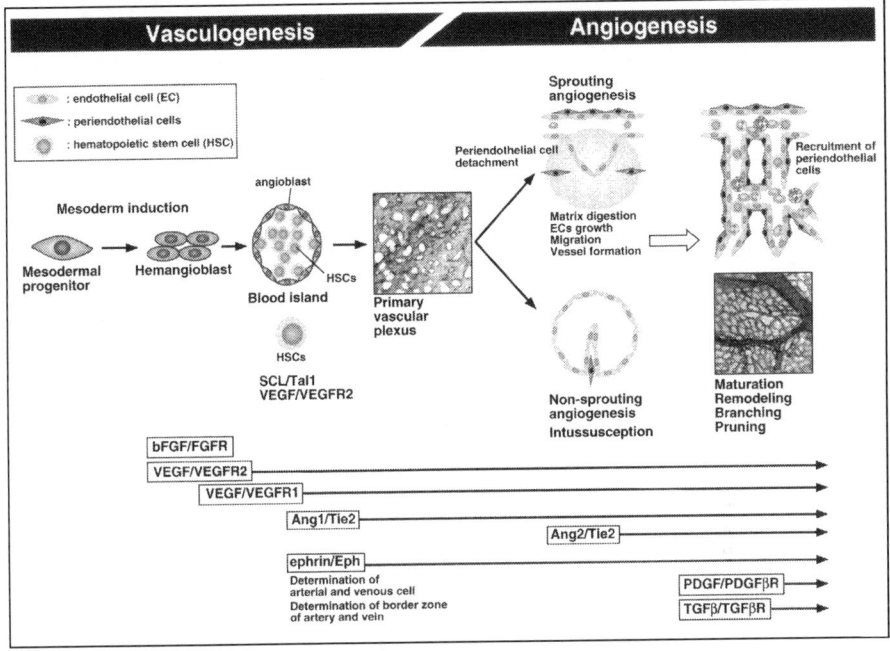

Figure 1. Vasculogenesis and angiogenesis. The early vascular plexus forms from mesoderm by differentiation of angioblasts that subsequently generate primitive blood vessels. The molecular mechanism responsible for this process is termed vasculogenesis. At this stage, VEGFs and their cognate receptors are the principal regulators of differentiation of angioblasts. After the primary vascular plexus is formed, more ECs are generated that can form new capillaries by sprouting or splitting from their vessel of origin in a process known as angiogenesis. There are at least two types of angiogenesis: true sprouting of capillaries from pre-existing vessels and non-sprouting angiogenesis or intussusception. At this stage, the Ang/Tie2 system plays important roles. At the final step of this stage, ECs need to be stabilized and matured. At this step, PDGF-BB recruits periendothelial cells, whereas Ang1 and TGF-β1 stabilize the nascent vessel.

known as angiogenesis.[1] Historically, the term angiogenesis was first used to describe the growth of endothelial sprouts from preexisting postcapillary venules. Recently, the term has been used to denote the growth and remodeling process of the primitive network into a complex network. The process of angiogenesis consists of sprouting, branching, fusion, intussusception, or pruning of ECs.[2] Repeated regression and rebirth of blood vessels is called remodeling, which includes detachment, migration, and adhesion of ECs. Analyses of gene knock-out mice (as summarized in Table 1) indicate that vascular endothelial growth factor (VEGF), angiopoietins, and ephrins play critical roles in development of blood vessels.

Vasculogenesis

Early vascular plexus forms from the mesoderm by differentiation of angioblasts (ECs that have not yet formed a lumen), which subsequently generates primitive blood vessels (Fig. 1). Mesoderm-inducing factors of the fibroblast growth factor family are crucial in inducing paraxial and lateral plate mesoderm to form angioblasts and HCs.[3] Existence of a bipotential precursor for these cell types (the so-called hemangioblast) is suggested by defects in both HCs and ECs seen in embryos lacking the vascular endothelial growth factor receptor-2 (VEGFR2)[4] (Table 1). There are six characterized VEGF homologues in mammals (VEGF-A through VEGF-E,

Table 1. Phenotypes of VEGF/VEGFR, Ang/Tie and ephrin/Eph deficient mice

Gene Knockout	Time of Death	Stage of Vessel Development	Causes of Lethality
VEGF-A(+/-)	E11.5	vasculogenesis (angiogenesis)	reduced red blood cells; defective heart and aorta formation, defective sprouting
VEGF-A(+/-)	E10.5	vasculogenesis	absent dorsal aorta; defective endothelial cell development
VEGFR1	E8.5-E9.5	vasculogenesis	failure of endothelial cell formation
VEGFR2	E8.5-9.5	vasculogenesis	excess endothelial cells form abnormal vessel structure entering vessel lumens
VEGFR3	E10.5-E12	vasculogenesis	defective vessel remodeling and organization; irregular large vessels with defective lumens
Neuropilin-1	E10.5	angiogenesis	anomalies in great vessels and heart outflow tracts, impairment in neural and pericardial vascularization
Ang1	E10.5	angiogenesis	defective vessel remodeling, organization, and sprouting; heart trabeculation defects
Ang2	E12.5-P1	maturity	poor vessel in integrity, edema, and hemorrhage
Tie1	E13.5-P1	maturity	poor vessel integrity, edema, and hemorrhage
Tie2	E10.5	angiogenesis	defective vessel remodeling, organization and sprouting; heart trabeculation defects
ephrin-B2	E10.5	vasculogenesis(angiogenesis)	some defective vessel primordia, defective vessel remodeling, organization and sprouting; heart trabeculation defects
EphB2/EphB3	E10.5(-30%)	vasculogenesis(angiogenesis)	some defective vessels primordia, defective vessel remodeling, organization and sprouting; heart trabeculation defects
EphB4	E10.5	vasculogenesis(angiogenesis)	defective cardiovascular development, defective peripheral angiogenesis, retardation or arrest of cardiac morphogenesis

VEGF/VEGFR mutants exhibit defects primarily in early stages of vasculogenesis, and accordingly, these embryos die at early developmental stages. By contrast, *Tie2* and *Ang1* knockout embryos die at later stages and exhibit similar defects; these mice exhibit normal vasculogenesis and perturbed angiogenesis. Deletion of *Ang2* results in lethality after the vascular system has undergone both vasculogenesis and angiogenesis. *EphB2/B3* and *ephrin-B2* mutant mice die primarily due to defects in angiogenesis similar to those seen in *Tie2* and *Ang1* knockout embryos, although some defects in vasculogenesis have also been reported. *EphB4* mutant mice have a symmetric phenotype with ephrin-B2 knockout embryos and die at E10.5.

and PlGF), and they display differential interactions with three related receptor tyrosine kinases (VEGFR1/Flt-1, VEGFR2/Flk-1/KDR, and VEGFR3/Flt-4) as summarized in Figure 2. Ancillary receptor components such as neuropilins also appear to interact these receptor complexes. The expression of VEGFR1 and VEGFR2 is restricted largely to the vascular endothelium, which accounts for the specificity of action of this growth factor family. Interestingly, although VEGFR3 is restricted largely to lymphatic endothelium,[5] mice lacking the gene encoding VEGFR3 display early embryonic lethality due to defects in the organization of large vessels prior to the emergence of lymphatics[6] (Table 1). The roles of VEGF-A and its receptors, VEGFR1 and VEGFR2, have been characterized in the greatest detail. In vitro, VEGF-A can induce endothelial cell proliferation as well as migratory and sprouting activity, and to promote formation of tubule-like structures by endothelial cells; these effects are mediated largely by VEGFR2. Consistent with these in vitro actions, VEGF-A and VEGFR2 are absolutely critical for the earliest stages of vasculogenesis in vivo, as blood islands, endothelial cells (Fig. 1), and major vessel tubes fail to develop in appreciable numbers in embryos lacking either VEGF-A or VEGFR-2.[8,47] After differentiation of ECs, VEGFR2 is downregulated in HCs but not in ECs. The other VEGF receptor, VEGFR1 (Flt-1), regulates vasculogenesis, since mice lacking VEGFR1 produce angioblasts, but their assembly into functional blood vessels is impaired[9] (Table 1). VEGF-A functions not only in the very initial phases of vasculogenesis, but in later stages of vasculogenesis such as sprouting and branching, as well as in maintaining vessel survival[7,8] (Fig.1). Threshold VEGF levels are required to maintain angioblast differentiation, as mice lacking one copy of the VEGF gene die in utero, with aberrant blood vessel formation in the yolk sac and embryo.[6,7] Intraembryonic angioblasts derived from the splanchnopleura can produce hematopoietic cells, unlike those derived from the somatopleuric mesoderm,[10] indicating that subtypes of angioblasts may exist. Little is known about the mechanisms governing endothelial cell fate: Ets-1, Hex, Vezf1, Hox, GATA family members, and basic helix-loop-helix factors and inhibitors of bHLH proteins may be involved in EC development.[11] Whether ECs become integrated into arteries or veins is mediated by the bHLH transcription factor gridlock at the angioblast stage, and subsequently, by members of the Eph-ephrin family, whose signals are also involved in guidance of axons and repulsion of neurons.[12]

Angiogenesis

The primary plexus is then remodeled by a process referred to as angiogenesis,[3] which includes the sprouting, branching, pruning and differential growth of vessels to form the more mature appearing vascular patterns, as seen in adult organisms. This latter phase of vascular development also requires differential recruitment of associated supporting cells, such as smooth muscle cells and pericytes, as well as fibroblasts, to different segments of vasculature.[13,14] There are at least two different types of angiogenesis. One is the true sprouting of capillaries from pre-existing vessels, and the other is non-sprouting angiogenesis or intussusception (Fig. 1).

Sprouting angiogenesis occurs both in the yolk sac and the embryo (most frequently during later organogenesis, particularly in the brain). Proteolytic degradation of extracellular matrix is followed by chemotactic migration and proliferation of ECs, lumen formation and functional maturation of endothelium.

Non-sprouting angiogenesis is a process of splitting pre-existing vessels by trans capillary pillars or posts of extracellular matrix,[15] first described in the embryonic lung.[16] Concurrent sprouting and non-sprouting angiogenesis are required for the vascularization of organs or tissues during development. In vivo, non-sprouting angiogenesis can occur by proliferation of ECs inside a vessel, producing a wide lumen that can be split by trans capillary pillars, or fusion and splitting of capillaries. The type of angiogenesis occurring in a given organ or tissue may depend on the number of vessels already present when the organ starts to grow rapidly. For example, sprouting angiogenesis occurs in the brain anlage that does not contain angioblasts, whereas non-sprouting angiogenesis predominates in the lung, which contains intrinsic EPCs (endothelial precursor cells) and is initially vascularized by vasculogenesis.[17]

For migration of ECs, it is necessary for ECs to release inter-endothelial cell contacts and relieve periendothelial cell (smooth muscle cells and pericytes) support; that is, mature vessels need to become destabilized.[2] Angiopoietin-2 (Ang2) might be involved in detaching smooth muscle cells and loosening the matrix in blood vessels.[12,18] Interestingly, Ang2 is required for recruitment of smooth muscle cells required for the stability and proper function of the collecting lymphatic vessels.[19] Proteinases of the plasminogen activator, matrix metalloproteinase (MMP), chymase or heparanase families modulate angiogenesis by degrading matrix molecules and activating or liberating growth factors (bFGF, VEGF, and IGF-1) sequestered within the extracellular matrix.[20] The exposed ECs proliferate and sprout from their resident site, eventually forming a new vessel. The loosening of cell-cell contacts allows the fusion of capillaries to form wider vessels, the arteries and veins. Eventually, mature capillary networks form and are stabilized by TGF-β which strengthens the extracellular matrix (Fig. 1). Angiopoietin-1 (Ang1) and platelet-derived growth factor are necessary for the recruitment of the pericytes that contribute to the mechanical flexibility of the capillary wall.[14]

Targeted mutations in mice have revealed some mechanisms responsible for angiogenesis. Early embryonic vessels (the vascular plexus formed by vasculogenesis) typically contain plump ECs forming wide lumina without a correspondingly thick vascular wall that may reflect VEGF's ability to induce formation of larger vessels by enhancing fusion of capillaries. Indeed, VEGF-deficient embryos lack the dorsal aorta.[7,21] The receptor tyrosine kinase Tie2 and its ligand Ang1 (Fig. 2) are also necessary for sprouting of ECs. Ang1 is a natural inhibitor of vascular permeability, tightening preexisting vessels. Ang1 phosphorylates Tie2, is chemotactic for ECs, induces sprouting and potentiates VEGF, but fails to induce proliferation of ECs.[12,22] In contrast to VEGF, Ang1 itself does not initiate endothelial network organization, but stabilizes networks initiated by VEGF, presumably by stimulating the interaction between endothelial and surrounding periendothelial cells. These findings indicate that Ang1 acts at later stages than VEGF.[12,22] In mice lacking either Ang1 or Tie2, early stages of VEGF-dependent vascular development (vasculogenesis) occur normally, resulting in the formation of a primitive vasculature.[23-25] However, remodeling and stabilization of this primitive vasculature is severely perturbed, leading to death around E10.5;[13,24,25] defects are particularly evident in the capillary plexi of the yolk sac and head (Table I). These defects likely result from disruptions in Tie2-Ang1 mediated interactions between ECs and surrounding periendothelial cells that produce Ang1. Another major defect in mice lacking Ang1 or Tie2 is in heart development (Table I). The transgenic overexpression of Ang1 in the skin leads to striking hypervascularization, presumably by promoting vascular remodeling events and perhaps by decreasing normal vascular pruning.[22] Transgenic overexpression of Ang2 in blood vessels during embryogenesis leads to a lethal phenotype reminiscent of that seen in Ang1 or Tie2 knockouts, with severe disruptions in vascular development.[18,26] Examination of angiopoietin expression patterns in vivo suggests an interesting role for Ang2 at vascular remodeling sites. Whereas Ang1 is expressed widely in normal adult tissues, consistent with a role of a continuously required stabilization of the vasculature, Ang2 is highly expressed only at sites of vascular remodeling in the adult, notably in the female reproductive tract.[18] Ang2 plays a facilitative role at sites of vascular remodeling in the adult by blocking a constitutive stabilizing action of Ang1, allowing the vessels to revert to a more plastic and unstable state[18] (Fig. 1).

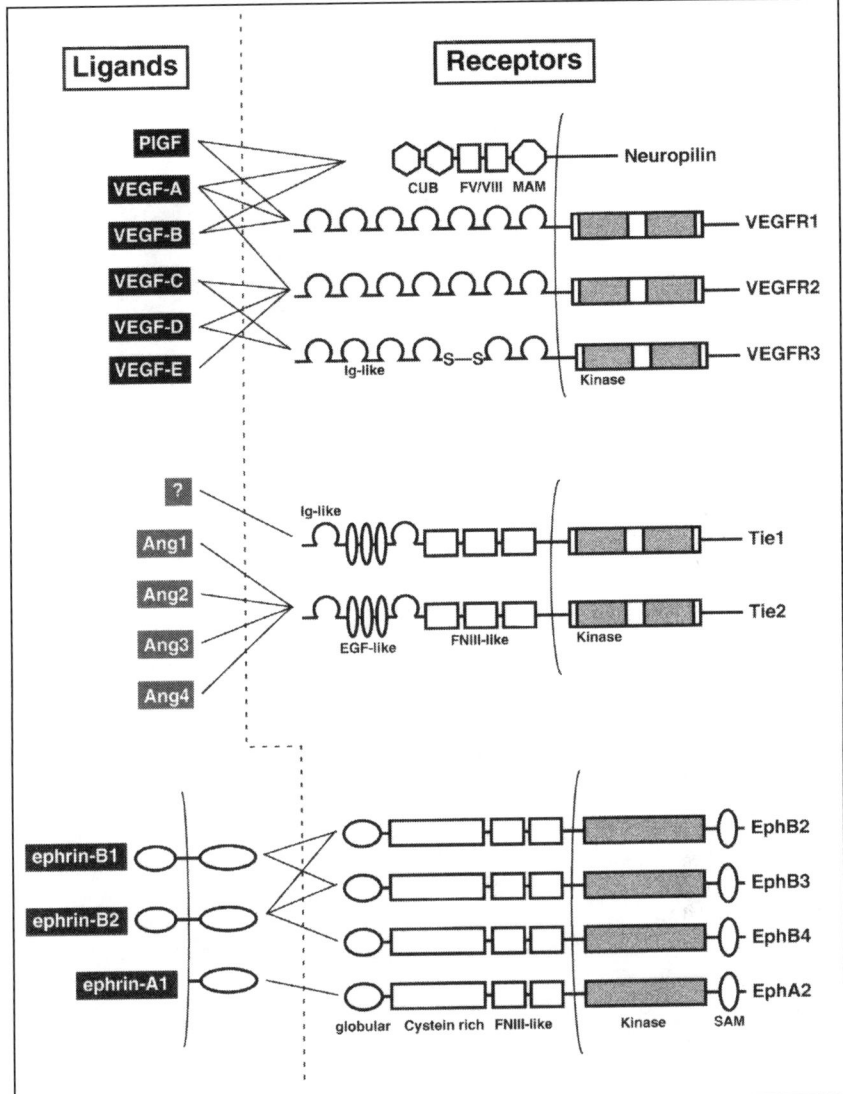

Figure 2. Lines indicate documented interactions among ligands with their receptors. Receptor and ligand structures are color coded to indicate their domains of expression. VEGFR1, VEGFR2, Tie1, Tie2, EphA2, and ephrin-B1 expressed on both arteries and veins. ephrin-B2 expressed on arteries. EphB3 and EphB4 are found on veins. VEGFR3 expressed on lymphatic vessels. PlGF, VEGF-A-D, Angiopoietin-1-4, and EphB2 expressed in tissues surrounding blood vessels.

Eph-Ephrin

The Eph receptor family consists of at least 14 members, and eight ephrin ligands for these receptors have been described to date.[27] Eph receptors and ligands are broadly divided into two subclasses, A and B, based on structural homologies and binding specificities, with a great deal of redundancy within a subclass in terms of receptor/ligand binding specificities.[28,29] The ephrin-B ligands are transmembrane proteins that preferentially bind to receptors of the

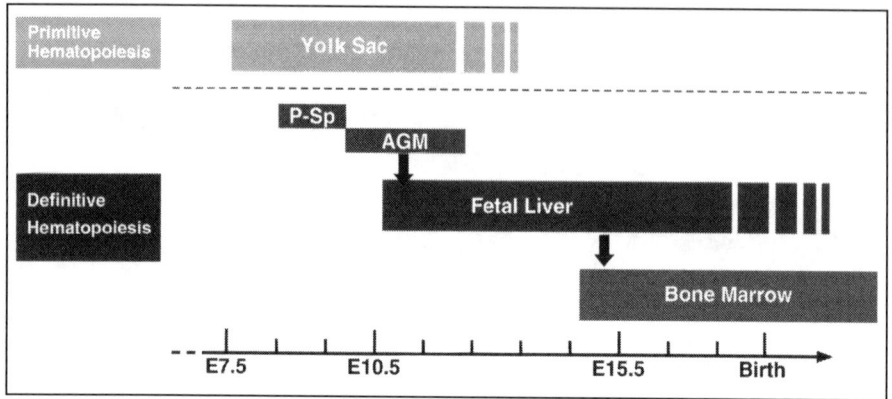

Figure 3. Sites of hematopoiesis. Hematopoiesis begins in the yolk sac at E7.5; subsequently, it shifts to the fetal liver and then to the spleen and bone marrow. Although, hematopoietic progenitors can be detected in the yolk sac as early as E7.5-8.5, these progenitors cannot contribute to definitive hematopoiesis (primitive hematopoiesis). Definitive hematopoiesis develops from the P-Sp region as early as E8.5. By E10.5, HSCs originate in the AGM. HSCs from this region colonize the fetal liver and then shift to the spleen and bone marrow.

Eph-B subclass. Unlike ligands for other receptor tyrosine kinases, the ephrins cannot act as soluble mediators but must be membrane-bound to activate their receptors (membrane linkage seemingly serves to cluster the ligands, and this clustering is required for activity).[30] Although in contrast to VEGF and angiopoietins, ephrins were initially characterized by their roles in axon guidance and neuronal patterning, and currently several findings suggest roles for the Eph family in the vasculature.[31]

Ephrin-B2 and EphB4 display remarkably reciprocal expression patterns during vascular development, with ephrin-B2 marking future arterial but not venous ECs while EphB4 marks the endothelium of primordial venous vessels.[32] These patterns suggest that ephrin-B2 and EphB4 function to establish arterial versus venous identity, perhaps in fusing arterial and venous vessels at their junctions, and suggest that bidirectional signaling occurs between such cells. In addition, the recent observation of vascular defects in ephrin-B2 and EphB4 deficient mice[32,33] suggests that interaction between the ephrin-B2 ligand and its cognate EphB4 receptor defines the boundaries of arterial-venous domains.[32,33] Subsequent work demonstrating expression of ephrin-B2 and cognate EphB receptors in mesenchymal cells adjacent to ECs suggests an Eph/ephrin-B2 interaction at the endothelial-mesenchymal contact zone.[34]

Development of HCs

During embryogenesis, development of the hematopoietic system occurs at various anatomical sites, including the extraembryonic yolk sac, fetal liver, spleen, and finally, bone marrow (Fig. 3). There are two discrete anatomic origins of hematopoietic activity, one extraembryonic and one intraembryonic.[35] In mouse embryogenesis, hematopoiesis begins in the yolk sac at E7.5; subsequently, it shifts to the fetal liver and then to the spleen and bone marrow[36] Hematopoiesis prior to formation of fetal liver is known as primitive hematopoiesis and is distinguished from adult-type definitive hematopoiesis by specific expression of embryonic-type globin in nucleated erythrocytes. During the primitive stage, groups of mesodermal cells aggregate in the developing yolk sac to form the blood islands. Cells at the periphery of these aggregates differentiate into angioblasts, while cells in the interior become primitive HCs.[37] Although

Table 2. Phenotypes of mouse mutant in molecules participating in hematopoiesis

Gene Knockout	Primitive Hematopoiesis	Definitive Hematopoiesis	Other Disorders
VEGFR2	X	X	deficit of endothelial cells
SCL/Tal-1	X	X	abnormal blood vessel formation in yolk sac
GATA-2	YS progenitor are modestly reduced	X	
Tie2	O	X	abnormal vascular network formation; heart trabeculation defects
AML1	O	X	poor branching of the cardial vein in head; poor network formation in pericardium
GATA-3	O	X	abnormalities in the nervous system
c-myb	O	X	

O: normal   X: deficit   YS: yolk sac

These genes are divided into two groups. One group results in abnormalities seen in not only hematopoiesis but in vasculo-angiogesis (VEGFR2, SCl/Tal-1, Tie2). The other group results in abnormalities seen primarily in definitive hematopoiesis (GATA-2, AML1, GATA-3, c-myb). Disorganized EC development and hemorrhage seen in AML1 mutant embryos are secondary effects due to a lack of definitive HSCs.

committed hematopoietic progenitors are detected in the yolk sac as early as E7.5-8.5, these progenitors have limited potential. Colony forming cells in spleen (CFU-S) and long-term repopulating hematopoietic stem cells (LTR-HSCs) are absent before the circulation is established. In the mouse embryo, LTR-HSCs develop from intraembryonic mesodermal region containing a para-aortic splanchnopleural mesoderm (P-Sp) as early as E7.5 in the absence of any yolk sac circulation.[38] By E10.5, LTR-HSCs originate in the embryo proper in the dorsal aorta, genital ridge/gonads and pro/mesonephros region (aorta-gonad-mesonephros: AGM), which is lineally related to the earlier P-Sp region and has been shown to harbor adult-type multipotent hematopoietic progenitors and pluripotential LTR-HSCs.[38-41] HSCs from this region are then presumed to colonize both yolk sac and fetal liver, where they give rise to definitive hematopoietic precursors after E12.5.[35] In contrast, recent studies indicate that the yolk sac HSCs at E9.0 and E10.0 can contribute to definitive hematopoiesis when busulfan treated newborn mice are used in place of irradiated adult mice as recipients for hematopoietic transplantation (see chapter by Palis).[42,43]

Regulation of Hematopoiesis

VEGFR2 and SCL/Tal1 are indispensable for not only primitive and definitive hematopoiesis but also for vasculo-angiogenesis. VEGFR2 is required for the development of hematopoietic and endothelial lineages in the early embryo. VEGFR2$^{-/-}$ mice lack both mature ECs and HCs as a result of failure to generate yolk sac blood islands during the primitive streak stage.[4] The defect in VEGFR2 deficient mice is thought to result from the inability of VEGFR2$^{-/-}$ mesodermal cells and/or hemangioblasts to migrate to the correct location to form blood islands, suggesting that VEGFR2 is required for movement of cells from the posterior primitive streak to the yolk sac and intraembryonic sites of early hematopoiesis. In addition, VEGF,

VEGFR1, and VEGFR2 are expressed on HSCs, and VEGF-dependent internal autocrine loops controls survival of HSCs.[44]

SCL/Tal1, which was originally identified through its translocation in acute T-cell lymphoblastic leukemia,[45-47] encodes a basic helix-loop-helix transcription factor expressed specifically in HCs,[48,49] vascular endothelium,[50,51] and the developing brain.[52] Targeted mutation of SCL/Tal1 indicates that it is indispensable to the development of both primitive and definitive hematopoiesis such as suitable establishment of the primary capillary plexus in the yolk sac.[53-56] The zebrafish mutant *cloche* affects both blood and endothelial differentiation.[57] Because of the loss of endocardium, the *cloche* mutant has enlarged cardiac chambers evident 26 hr post-fertilization (hpf).[58] *cloche* homozygotes exhibit nearly undetectable expression levels of GATA-1 and VEGFR2 and complete loss of Tie1.[58] Correspondingly, it has been shown that *cloche* mutants fail to produce differentiated blood and angioblasts. Genetic analysis demonstrates that the *cloche* mutation is not linked to the *SCL* locus. Forced expression of *SCL* in *cloche* embryos rescues the blood and vascular defects, suggesting that *SCL* acts downstream of *cloche* to specify hematopoietic and vascular differentiation.[59]

TIE2 and Ang1 are also important for definitive hematopoiesis. Tie2 deficient mice show abnormal vascular network formation. An activating mutation in Tie2 causes vascular dysmorphogenesis in human.[60] Although Ang1 alone does not promote proliferation in HCs, proliferation is promoted if Ang1 is added to a culture with along with stem cell factor. Furthermore, Tie2-Ang1 signaling promotes integrin dependent cell adhesion to fibronectin or collagen.[61,62]

Transcription factors such as GATA-2,[63] AML1,[64,65] GATA-3,[66] c-myb[67] are important effectors for definitive hematopoiesis. The phenotypes of mice mutant for these molecules are summarized in Table II.

Hemangioblasts: Common Progenitors of HCs and ECs

Hematopoiesis is closely related to angiogenesis, indicating the existence of common progenitors, hemangioblasts,[68] which are able to differentiate into both HSCs and endothelial cells. Hematopoietic and endothelial lineages express a number of genes in common, including CD34,[69] VEGFR2,[68-71] Flt-1,[72] Tie2,[61] SCL/Tal-1,[55] GATA-2[73] and PECAM-1.[74] The fact that these lineages co-express these genes, many of which encode growth factor receptors or transcription factors, is consistent not only with the notion that they share a common precursor, but also suggests that similar molecular programs and growth regulatory mechanisms are utilized as their development progresses. Histological studies in chick,[75] mouse[76] and human[76] have demonstrated the presence of clusters of HCs in close association with, and often adhering to, endothelial cells on the ventral surface (floor) of the aorta. The observation that the appearance of these clusters coincides with the onset of hematopoietic activity at this site suggests that they contain developing stem cells and precursors. As with the ECs in this region of the aorta, HCs that form these clusters also appear to be derived from P-Sp. In addition to the aorta, hematopoietic clusters have also been identified in the vitelline and umbilical arteries, indicating that intraembryonic hematopoietic development may be associated with the major arterial region of the embryo.[76,78,79] The structure of these intraembryonic clusters differs from that of the yolk sac blood islands, but the close association of the hematopoietic and endothelial lineages in the arterial region of the embryo has led to speculation that they could be sites of hemangioblast development. These intraembryonic hemangioblasts would likely differ from their counterparts in the yolk sac in that they should be restricted to the definitive hematopoietic system, as there is no detectable primitive erythroid lineage cell outside the yolk sac. Gene targeting studies in mice demonstrate that VEGFR2,[4,80] SCL/Tal-1[53,54] and TGF-β[81] are

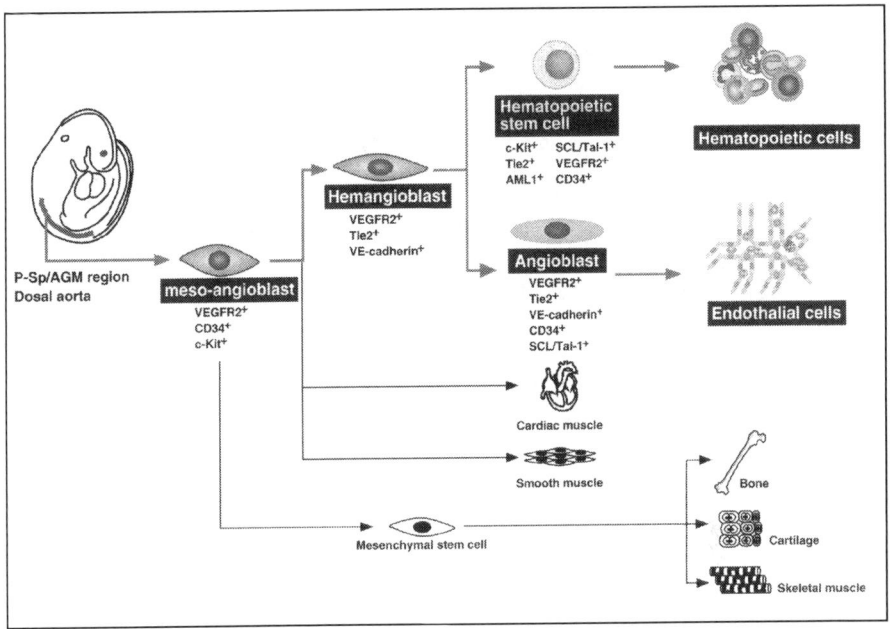

Figure 4. Differentiation of meso-angioblast and hemangioblast. Stem cells, called meso-angioblasts, exist in the dorsal aorta, particularly in the AGM region. Meso-angioblasts originate in the endothelial and sub-endothelial cells, and can differentiate into most mesodermal lineages (vessels, blood, cartilage, bone, smooth muscle, skeletal muscle, cardiac muscle). It is thought that meso-angioblasts contributes to mesodermal development after birth, and hemangioblasts may be contained in meso-angioblast populations. In addition, hematopoietic and endothelial lineages express a number of genes in common, such as VEGFR2, Tie2, CD34, SCL/Tal-1.

essential for normal development and growth of both lineages. As noted above *cloche* mutations in zebrafish show disrupted embryonic development of both hematopoietic lineages and endocardium.[57]

A direct approach to the identification of the hemangioblast utilizes surface markers as a means to isolate it. Several studies have employed this strategy, providing further evidence for the existence of hemangioblasts. However, none has yet identified a single cell that can give rise to both the hematopoietic and endothelial lineages. Eichmann et al[69] sorted VEGFR2+ cells from the mesoderm of the early chick embryo and found that this VEGFR2+ cells could differentiate into both HCs and ECs, suggesting that VEGFR2+ cells contain hemangioblasts. Analyses of single sorted cells, however, demonstrated that single VEGFR2+ cells could generate either HCs or ECs, but not both, as different conditions were required for the development of these two lineages. Consequently, it is difficult to determine from this study whether VEGFR2+ cells contain a mixture of lineage-restricted precursors or a multipotential hemangioblast. Nishikawa et al[82] isolated mouse yolk sac and P-Sp/AGM cells based on the expression of VE-cadherin and demonstrated multilineage hematopoietic potential in these VE-cadherin+ cells. Since VE-cadherin was regarded as a marker of ECs, these findings were interpreted as evidence that HCs can develop from a specific subpopulation of ECs with hemangioblastic potential. Hamaguchi et al[83] isolated Tie2+ precursors from the AGM region at E10.5 which could differentiate into HCs as well as cells with endothelial characteristics as defined by PECAM-1 expression. These findings suggest that these AGM-derived Tie2[+] cells contain hemangioblasts.

In a more recent study, Minasi et al[84] reported that the dorsal aorta contains progenitors for multiple mesodermal tissues. These aorta associated mesodermal progenitor cells "called meso-angioblasts" can differentiate into most mesodermal lineages, such as blood vessels, blood, bone, cartilage, smooth, skeletal and cardiac muscles, in vivo and in vitro. These cells originate in endothelial and sub-endothelial cells of the embryonic dorsal aorta, especially the AGM region. When quail or mouse embryonic aorta were grafted into host chick embryos, donor aorta derived cells, initially incorporated into the host vessels, were later integrated into meso-dermal tissues.[84] Moreover, the expanded progeny from mouse dorsal aorta expressed hemato-poietic and endothelial cell markers (CD34, VEGFR2, and c-Kit) at both early and late passages, and maintained multipotency in culture or following transplantation into a chick embryo. It is possible that this vessel associated stem cell, the meso-angioblast, participates in postembryonic development of the mesodermal system organization and that population of meso-angioblasts contains subpopulations of hemangioblasts (Fig. 4).

Interaction between HSCs and ECs

As described above, intraembryonic hematopoietic development may be associated with the major arterial region. It suggests that ECs provide hematopoietic microenvironment and contribute to the development of HCs. Alternatively, it was recently reported that the HSC itself contributes to remodeling and angiogenesis of blood vessels during embryogenesis.[80]

Takakura et al[85] analyzed AML1-deficent mice to determine the interaction between HSCs and ECs. Disruption of AML1 leads to failure in development of definitive hematopoiesis and lethality at E12.5.[64,65] AML1[-/-] embryos exhibit hemorrhages in the ventricles of the central nervous system, in the vertebral canal and within the pericardial space and peritoneal cavity. Vascular branching and remodeling into large and small vessels occurs normally in the head region of both wild-type and mutants up to E11.5. However, the number of small capillaries seen in the hindbrain of AML1[-/-] embryos was less than that observed in wild-type mice; moreover, large vessels in the AML1[-/-] embryos contained fewer branches than those seen in wild-type mice. In mutant embryos, less branching of capillaries was observed in vessels of the pericardium and the vitelline artery of the yolk sac. Consistent with defective angiogenesis in the head, massive hemorrhages and aneurysms were observed in mutant embryos.

Recently, it has been reported that AML1 is expressed in ECs at sites where early HSCs emerge, such as the yolk sac, vitelline and umbilical arteries, and the dorsal aorta in the AGM region.[86] An autonomous effect in endothelial cells by disruption of the AML1 gene has been suggested to cause hemorrhage; however, AML1 expression in ECs is not observed in the brain and heart where severe abnormalities in angiogenesis are observed in mutant embryos. In order to examine angiogenic activity of the AML-1[-/-] mouse, P-Sp region of E9.5 mouse embryo is cocultured on OP9 stromal cells. In this culture, ECs form a sheet-like structure (vascular bed) and subsequently form a network (vascular network) in the periphery of the vascular sheet. Formation of the vascular network is seen in sites of HC proliferation. Compared with wild-type mice, proliferation of endothelial cells observed in AML1[-/-] mice was normal, while vascular network formation was decreased. Moreover, no HCs were generated in P-Sp cultures from AML1 mutant. Addition of HSCs to this cultivation system nearly rescued impaired angiogenesis in AML1 mutants (Fig. 5). Recovery of vascular network was seen in accordance with the region that HCs, especially HSCs, proliferated. This finding suggests that disorganized EC development and hemorrhage seen in AML1 mutant embryos are secondary effects due to a lack of definitive HSCs. Moreover, HSCs (Lin[-]c-Kit[+]Sca-1[+] cells in adult bone marrow, and

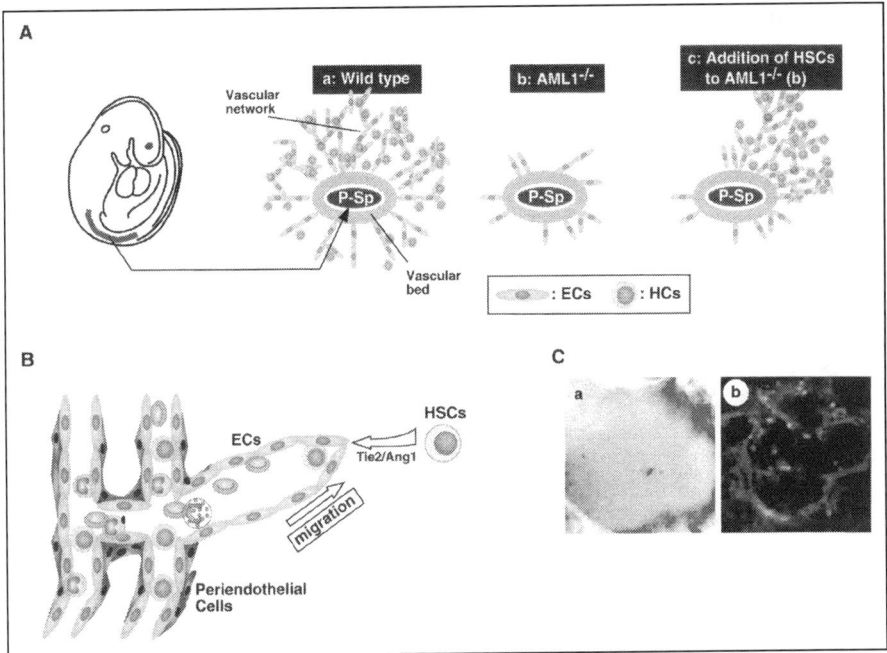

Figure 5. HSCs promote angiogenesis. A) Organ culture of P-Sp on OP9 Stromal cells. Vasculo-angiogenesis and hematopoiesis are observed. Formation of blood vessel networks is seen in the regions containing HCs, especially as HSCs increase (a). Cultivation of P-Sp derived from AML1 mutants. Vascular network formation and HC generation were not observed (b). HSCs of wild-type mice were added to cultures derived from AML1 mutants (b). Formation of the vascular network was guided in accordance with the region that HCs form colonies (c). B) Ang1 derived from HSCs promotes vessel sprouting into avascular areas. C) HSCs promote capillary sprouting in vivo. Matrigel containing HSCs was injected subcutaneously into adult mice. a) Appearance of dissected matrigel on day 4. b) Angiogenesis was guided in matrigel. Red: HSCs. Green; ECs.

CD45+c-Kit+CD34+ cells in embryo) produce Ang-1.[85] It is surmised that the rescue of vascular network by the addition of HSCs depends on Ang-1 secreted by HSCs. When Ang-1 was added to the AML1$^{-/-}$ mouse P-Sp culture, angiogenesis was recovered. Taken together this data suggests the following model: blood vessels migrate to a non-vascular field, HSCs migrate to that field and secrete Ang-1 and guide the chemotaxis of Tie2+ ECs (Fig. 5).

In summary, development of endothelial cells and hematopoietic cells are closely related to each other in embryonic stages, and both cell types interact with each other through extrinsic factors such as VEGFs and Angiopoietins.

References

1. Risau W. Differentiation of endothelium. FASEB J 1995; 9:926-993.
2. Carmerliet P. Mechanisms of angiogenesis and arteriogenesis. Nature. 2000; 6:389-395.
3. Risau W. Mechanisms of angiogenesis. Nature 1997; 386:671-674.
4. Shalaby F, Rossant J, Yamaguchi TP et al. Failure of blood-island formation and vasculogenesis in Flk-1-deficient mice. Nature 1995; 376:62-66.
5. Kukk E, Lymboussaki A, Taira S et al. VEGF-C receptor binding and pattern of expression with VEGFR-3 suggests a role in lymphatic vascular development. Development 1996; 122:3829-3837.

6. Dumont DJ, Jussila L, Taipale J et al. Cardiovascular failure in mouse embryos deficient in VEGF receptor-3. Science 1998; 282:946-949.

7. Carmeliet P, Ferreira V, Breier G et al. Abnormal blood vessel development and lethality in embryos lacking a single VEGF allele. Nature 1996; 380:435-439.

8. Ferrara N, Carver-Moore K, Chen H et al. Heterozygous embryonic lethality induced by targeted inactivation of the VEGF gene. Nature 1996; 380:439-442.

9. Fong GH, Rossant J, Gertsenstein M et al. Role of the Flt-1 receptor tyrosine kinase in regulating the assembly of vascular endothelium. Nature 1995; 376:66-70.

10. Pardanaud L, Luton D, Prigent M et al. Two distinct endothelial lineages in ontogeny, one of them related to hemopoiesis. Development 1996; 122:1363-1371.

11. Lyden D, Young AZ, Zagzag D et al. Id1 and Id3 are required for neurogenesis, angiogenesis and vascularization of tumour xenografts. Nature 1999; 401:670-677.

12. Gale NW, Yancopoulos GD. Growth factors acting via endothelial cell-specific receptor tyrosine kinases:VEGFs, Angiopoietins, and ephrins in vascular development. Genes Dev 1999; 13:1055-1066.

13. Folkman J, D'Amore PA. Blood vessel formation: What is its molecular basis? Cell 1996; 87:1153-1155.

14. Lindahl P, Johansson BR, Leveen P et al. Pericyte loss and microaneurysm formation in PDGF-B-deficient mice. Science 1997; 277:242-245.

15. Short RHD. Alveolar epithelium in relation to growth of the lung. Phil Trans R Soc Lond B 1950; 235:35-87.

16. Patan S, Haenni B, Burri PH. Implementation of intussusceptive microvascular growth in the chicken chorioallantoic membrane. 1. Pillar forming by folding of the capillary wall. Microvasc Res 1996; 51:80-98.

17. Pardanaud L, Yassine F, Dieterlin-Lievre F. Relationship between vasculogenesis, angiogenesis and hematopoiesis during avian ontogeny. Development 1989; 105:473-485.

18. Maisonpierre PC, Suri C, Jones PF et al. Angiopoietin-2, a natural antagonist for Tie2 that disrupts in vivo angiogenesis. Science 1997; 277:55-60.

19. Gale NW, Thurston G, Hackett SF et al. Angiopoietin-2 is required for postnatal angiogenesis and lymphatic patterning, and only the latter role is rescued by angiopoietin-1. Dev Cell 2002; 3:411-423.

20. Coussens LM, Raymond WW, Bergers G et al. Inflammatory mast cells up-regulate angiogenesis during squamous epithelial carcinogenesis. Genes Dev 1999; 13:1382-1397.

21. Drake CJ, Little CD. Exogeneous vascular endothelial growth-factor induces malformed and hyperfused vessels during embryonic neovascularization. Proc Natl Acad Sci USA 1995; 92:7657-7661.

22. Suri C, McClain J, Thurston G et al. Increased vascularization in mice overexpressing Angiopoietin-1. Science 1998; 282:468-471.

23. Dumont DJ, Gradwohl G, Fong GH et al. Dominant-negative and targeted null mutations in the endothelial receptor tyrosine kinase, tek, reveal a critical role in vasculogenesis of the embryo. Genes Dev 1994; 8:1897-1909.

24. Sato TN, Tozawa Y, Deutsch U et al. Distinct roles of the receptor tyrosine kinases Tie-1 and Tie-2 in blood vessel formation. Nature 1995; 376:70-74.

25. Suri C, Jones PF, Patan S et al. Requisite role of Angiopoietin-1, a ligand for the TIE2 receptor, during embryonic angiogenesis. Cell 1996; 87:1171-1180.

26. Hanahan D. Signaling vascular morphogenesis and maintenance. Science 1997; 277:48-50.

27. Yancopoulos GD, Klagsbrun M, Folkman J. Vasculogenesis, angiogenesis and growth factors: ephrins enter the fray at the border. Cell 1998; 93:661-664.

28. Gale NW, Holland SJ, Valenzuela DM et al. Eph receptors and ligands comprise two major specificity subclasses and are reciprocally compartmentalized during embryogenesis. Neuron 1996; 17:9-19.

29. Flanagan JG, Vanderhaeghen P. The ephrins and Eph receptors in neural development. Annu Rev Neurosci 1998; 21:309-345.
30. Davis S, Gale NW, Aldrich TH et al. Ligands for EPH-related receptor tyrosine kinases that require membrane attachment or clustering for activity. Science 1994; 266:816-819.
31. Stein E, Lane AA, Cerretti DP et al. Eph receptors discriminate specific ligand oligomers to determine alternative signaling complexes, attachment, and assembly responses. Genes Dev 1998; 12:667-678.
32. Wang HU, Chen ZF, Anderson DJ. Molecular distinction and angiogenic interaction between embryonic arteries and veins revealed by ephrin-B2 and its receptor Eph-B4. Cell 1998; 93:741-753.
33. Gerety SS, Wang HU, Chen ZF et al. Symmetrical mutant phenotypes of the receptor EphB4 and its specific transmembrane ligand ephrin-B2 in cardiovascular development. Mol Cell 1999; 4:403-414.
34. Adams RH, Wilkinson GA, Weiss C et al. Roles of ephrinB ligands and EphB receptors in cardiovascular development: Demarcation of arterial/venous domains, vascular morphogenesis, and sprouting angiogenesis. Genes Dev 1999; 13:295-306.
35. Dzierzak E, Medvinsky A. Mouse embryonic hematopoiesis. Trends Genet 1995; 11:359-366.
36. Johnson GR, Moore MAS. Role of stem cell migration in initiation of mouse foetal liver hematopoiesis. Nature 1975; 258:726-728.
37. Risau W. Embryonic angiogenesis factors. Pharmac Ther 1991; 51:371-376.
38. Cumano A, Diterien-Lievre F, Godin I. Lymphoid potential, Probed before circulation in mouse, is restricted to caudal intraembryonic splanchnopleure. Cell 1996; 86:907-916.
39. Medvinsky AL, Samoylina NL, Müller AM et al. An early pre-liver intra-embryonic source of CFU-S in the developing mouse. Nature 1993; 364:64-66.
40. Müller AM, Medvinsky A, Strouboulis J et al. Development of hematopoietic stem cell activity in the mouse embryo. Immunity 1994; 1:291-301.
41. Medvinsky A, Dzierzak E. Definitive hematopoiesis is autonomously initiated by the AGM region. Cell 1996; 86:897-906.
42. Yoder MC, Hiatt K, Dutt P et al. Characterization of definitive lymphohematopoietic stem cells in the day 9 murine yolk sac. Immunity 1997a; 7:335-344.
43. Yoder MC, Hiatt K, Mukherjee P. In vitro repopulating hematopoietic stem cells are present in the murine yolk sac at day 9.0 postcoitus. Proc Natl Acad Sci USA 1997b; 94:6776-6780.
44. Gerber HP, Malik AK, Solar GP et al. VEGF regulated haematopoietic stem cell survival by an internal autocrine loop mechanism. Nature 2002; 417:954-958.
45. Begley CG, Aplan PD, Davey MP et al. Chromosomal translocation in human leukemic stem-cell line disrupts the T-cell receptor delta-chain diversity region and results in a previously unreported fusion transcript. Proc Natl Acad Sci USA 1989; 86:2031-2035.
46. Finger LR, Kagen J, Christopher G et al. Involvement of the TCL5 gene on human chromosome 1 in T-cell leukemia and melanoma. Proc Natl Acad Sci USA 1989; 86:5039-5043.
47. Chen Q, Yang CY, Tsan JT et al. Coding sequence of the Tal-1 gene are disrupted by chromosome translocation in human T cell leukemia. J Exp Med 1990; 172:1403-1408.
48. Green AR, Salvaris E, Begley CG. Erythroid expression of the "helix-loop-helix" gene, SCL. Oncogene 1991; 6:475-479.
49. Visvader J, Begley CG, Adams JM. Differential expression of the LYL, SCL, E2A helix-loop-helix genes within the hematopoietic system. Oncogene 1991; 6:187-194.
50. Hwang L-Y, Siegelman M, Davis L et al. Expression of the TAL1 proto-oncogene in cultured endothelial cells and blood vessels of the spleen. Oncogene 1993; 8:3043-3046.
51. Kallianpur AR, Jordan JE, Brandt SJ. The SCL/TAL-1 gene is expressed in progenitors of both the hematopoietic and vascular systems during embryogenesis. Blood 1994; 83:1200-1208.
52. Green AR, Lints T, Visvader J et al. SCL is coexpressed with GATA-1 in haemopoietic cells but is also expressed in developing brain. Oncogene 1992; 7:653-660.
53. Robb L, Lyons I, Li R et al. Absence of yolk sac hematopoiesis from mice with a targeted disruption of the scl gene. Proc Natl Acad Sci USA 1995; 92:7075-7079.
54. Shivdasani RA, Mayer EL, Orkin SH. Absence of blood formation in mice lacking the T cell leukaemia protein tal-1/SCL. Nature 1995; 373:432-434.

55. Visvader JE, Fujiwara Y, Orkin SH. Unsuspected role for the T-cell leukemia protein SCL/tal-1 in vascular development. Genes Dev 1998; 12:473-479.
56. Elefanty AG, Begley CG, Hartley L et al. SCL expression in the mouse embryo detected with a targeted lacZ reporter gene demonstrates its localization to hematopoietic, vascular, and neural tissues. Blood 1999; 94:3754-3763.
57. Stainer DY, Weintein BM, Detrich III et al. cloche, an early acting zebrafish gene, is required by both the endothelial and hematopoietic lineages. Development 1995; 121:3141-3150.
58. Liao W, Bisgrove BW, Sawyer H et al. The zebrafish gene cloche acts upstream of a flk-1 homologue to regulate endothelial differentiation. Development 1997; 124:381-389.
59. Liao EC, Paw BH, Oates AC et al. SCL/Tal-1 transcription factor acts downstream of cloche to specify hematopoietic and vascular progenitors in zebrafish. Genes Dev 1998; 12:621-626.
60. Vikkula M, Boon LM, Carraway KL III et al. Vascular dysmorphogenesis caused by an activating mutation in the receptor tyrosine kinase TIE2. Cell 1996; 87:1181-1190.
61. Takakura N, Huang XL, Naruse T et al. Critical role of the TIE2 endothelial receptor in the development of definitive hematopoiesis. Immunity 1998; 9:677-686.
62. Sato A, Iwama A, Takakura N et al. Characterization of TEK receptor tyrosine kinase and its ligands, Angiopoietins, in human hematopoietic progenitor cells. Int Immuno 1998; 10:1217-1227.
63. Tsai FY, Keller G, Kuo FC et al. An early haematopoietic defect in mice lacking the transcription factor GATA-2. Nature 1994; 371:221-226.
64. Okuda T, van Deursen J, Hiebert SW et al. AML1, the target of multiple chromosomal translocations in human leukemia, is essential for normal fetal liver hematopoiesis. Cell 1996; 84:321-330.
65. Wang Q, Stacy T, Binder M et al. Disruption of the Cbfa2 gene causes necrosis and hemorrhaging in the central nervous system and blocks definitive hematopoiesis. Proc Natl Acad Sci USA 1996; 93:3444-3449.
66. Pandolfi PP, Roth ME, Karis A et al. Targeted disruption of the GATA3 gene causes severe abnormalities in the nervous system and in fetal liver haematopoiesis. Nat Genet 1995; 11:40-44.
67. Mucenski ML, McLain K, Kier AB et al. A functional c-myb gene is required for normal murine fetal hepatic hematopoiesis. Cell 1991; 65:677-689.
68. Eichmann A, Corbel C, Nataf V et al. Ligand-dependent development of the endothelial and hematopoietic lineages from embryonic mesodermal cells expressing vascular endothelial growth factor receptor 2. Proc Natl Acad Sci USA 1997; 94:5141-5146.
69. Young PE, Baumhueter S, Lasky LA. The sialomucin CD34 is expressed on hematopoietic cells and blood vessels during murine development. Blood 1995; 85:96-105.
70. Millauer B, Wizigmann-Voos S, Schnurch H et al. High affinity VEGF binding and developmental expression suggest Flk-1 as a major regulator of vasculogenesis and angiogenesis. Cell 1993; 72:835-846.
71. Kabrun N, Buhring HJ, Choi K et al. Flk-1 expression defines a population of early embryonic hematopoietic precursors. Development 1997; 124:2039-2048.
72. Fong GH, Klingensmith J, Wood CR et al. Regulation of flt-1 expression during mouse embryogenesis suggests a role in the establishment of vascular endothelium. Dev Dyn 1996; 207:1-10.
73. Orkin S. GATA-binding transcription factors in hematopoietic cells. Blood 1992; 80:575-581.
74. Watt SM, Gechmeissener SE, Bates PA. PECAM-1: Its expression and function as a cell adhesion molecule on hematopoietic and endothelial cells. Leuk Lymphoma 1995; 17:229-244.
75. Dieterlen-Lièvre F, Martin C. Diffuse intraembryonic hemopoiesis in normal and chimeric avian development. Dev Biol 1981; 88:180-191.
76. Garcia-Porrero JA, Godin IE, Dieterlen-Lièvre F. Potential intraembryonic hemogenic sites at pre-liver stages in the mouse. Anat Embryol 1995; 192:425-435.
77. Tavian M, Coulombel L, Luton D et al. Aorta-associated CD34$^+$ hematopoietic cells in the early human embryo. Blood 1996; 87:67-72.
78. Wood HB, May G, Healy L et al. CD34 expression patterns during early mouse development are related to models of blood vessel formation and reveal additional sites of hematopoiesis. Blood 1997; 90:2300-2311.
79. de Bruijn MF, Speck NA, Peeters MC et al. Definitive hematopoietic stem cells first develop within the major arterial regions of the mouse embryo. EMBO J 2000; 19:2465-2474.

80. Shalaby F, Ho J, Stanford WL et al. A requirement for Flk1 in primitive and definitive hemato-poiesis and vasculogenesis. Cell 1997; 89:981-990.

81. Dickson MC, Martin JS, Cousins FM et al. Defective haematopoiesis and vasculogenesis in trans-forming growth factor-β1 knock out mice. Development 1995; 121:1845-1854.

82. Nishikawa S-I, Nishikawa S, Kawamoto H et al. In vitro generation of lymphohematopoietic cells from endothelial cells purified from murine embryos. Immunity 1998; 8:761-769.

83. Hamaguchi I, Huang XL, Takakura N et al. In vitro hematopoietic and endothelial cell develop-ment from cells expressing TEK receptor in murine aorta-gonad-mesonephros region. Blood 1999; 93:1549-1556.

84. Minsai MG, Riminucci M, De Angelis L et al. The meso-angioblast: A multipotent, self-renewing cell that originated from the dorsal aorta and differentiates into most mesodermal tissues. Develop-ment 2002; 129:2773-2783.

85. Takakura N, Watanabe T, Suenobu S et al. A role for hematopoietic stem cells in promoting angiogenesis. Cell 2000; 102:199-209.

86. North T, Gu TL, Stacy T et al. Cbfa2 is required for the formation of intra-aortic hematopoietic clusters. Development 1999; 126:2563-2575.

Origin and Fate of Hematopoietic Precursors in the Early Mouse Embryo

Julien Yuan Bertrand, Alexandra Manaia, Jeanne Van Celst, Ana Cumano and Isabelle Godin

Introduction

The hematopoietic system comprises a large array of differentiated cells (lymphocytes, myeloid cells, erythrocytes, etc...) with a limited half-life. The hematopoietic compartment thus depends on a continuous renewal throughout life. In adult mammals, mature blood cells are constantly generated in the bone marrow, from multipotent precursors endowed with self-renewal capacity: the hematopoietic stem cells (HSC), which have been considered to arise from a pool generated during ontogeny. Until recently, this model had not been tested experimentally. The existence of a lineage relationship between HSC present in the adult bone marrow and those generated during ontogeny recently receive support from experiments involving time-induced recombination event in HSC from early hematopoietic site, resulting in the activation of the gene encoding LacZ.[1]

The fetal liver (FL) was the first embryonic site identified as containing HSC, from 11 days post coitus (dpc) until the perinatal period.[2] The FL is the major hematopoietic organ in the embryo and contributes to neonatal hematopoiesis until the two first weeks after parturition.[3] At about the same stage, the thymus is already involved in the generation of T cells (from 10-11 dpc onwards). The fetal spleen, which seems to have only a minor contribution to embryonic hematopoiesis, has also been shown to contain HSC from 14 dpc until birth.[4,5] The fetal bone marrow is thought to harbor HSC from 15-16 dpc.[6]

Studies performed in the mouse and chicken embryos have shown that hematopoietic organs (FL, thymus, fetal spleen and also the bone marrow) do not produce de novo their own hematopoietic precursors, but have to be colonized by extrinsic hematopoietic cells.[5,7,8]

The only tissue where de novo hematopoiesis occurs is the Yolk Sac (YS). Indeed, the first blood cells can be detected in this site at 7.5 dpc, soon after the initiation of gastrulation. The early emergence of hematopoietic cells in the YS, together with their de novo generation have been for long the arguments in favor of the origin of HSC in the YS. It was then widely accepted that YS-derived HSC migrate to the FL, from where they colonize the other hematopoietic organs, including the bone marrow where they will reside throughout life.

This hematopoietic development scheme held until it was shown, in the avian model, that the pool of definitive HSC originates in the embryo proper (See the article by Jaffredo et al, this issue). Using the quail-chick chimeras, where a quail embryo is grafted onto a chick YS, it was shown that the first wave of blood cells derived from the YS is progressively replaced by

Hematopoietic Stem Cell Development, edited by Isabelle Godin and Ana Cumano.
©2006 Eurekah.com and Kluwer Academic / Plenum Publishers.

intra-embryonic (quail)-derived cells.[9] These data pointed to the embryo proper and not to the YS as the origin of definitive HSC.

In this review, we will examine the extra- and intra-embryonic origin of hematopoietic precursors in the early mouse embryo and the features and respective contribution of both sets of precursors to the developing hematopoietic system. We first show evidence that point to an independent generation of YS and intra-embryonic hematopoietic precursors. Then, through the analysis of phenotype and potential of the first extra- and intra-embryonic hematopoietic precursors, correlated to the analysis of gene expression patterns in these two sites, we focus on the origin of these cells in the embryo.

Extra- and Intraembryonic Contribution to Hematopoietic Development

The Embryo Proper as a Source of Multipotent Progenitors/HSC: The Discovery

The existence of lymphoid/multipotent progenitors in the embryo proper was independently reported by four groups in the early nineties. Ogawa et al first showed that B cell progenitors (detected in vitro) appeared earlier in the embryonic compartment than in the YS, respectively at 9.5 dpc and 10 dpc.[6] Cumano et al established by limiting dilution assays, the concomitant appearance of B cell precursors in the YS and embryo as soon as the 10-12 somite (S)-stage, at 8.5 dpc. Moreover, this study pointed to the AA4.1$^+$ Sca-1$^-$ population as containing the lymphoid precursors, although a precise localization of these precursors in both sites was not achieved.[10]

The presence in the intra-embryonic compartment of precursors displaying a B-lymphoid potential was corroborated by Godin et al., who characterized the intra-embryonic Para-aortic Splanchnopleura (P-Sp) as a source of B-1a cells at 9.5 dpc.[11] At this time, the P-Sp consists of the mesodermal territory that comprises the dorsal aortae (paired at this stage). In this study, P-Sp were grafted under the kidney capsule of host SCID recipients (deficient for B and T lymphocytes), which were repopulated with B-1a cells in the peritoneal cavity. Using a similar protocol, it was impossible to observe a similar reconstitution from the YS. Thus, although the reconstitution was specific to the B lineage, it clearly shows the existence of precursors displaying a lymphoid potential in the intra-embryonic site, the P-Sp (11).

The first evidence for the existence of more immature hematopoietic precursors in the mouse intra-embryonic compartment arose from CFU-S assay (then considered capable to detect the most immature hematopoietic precursors, including HSC,[12] performed at 10 dpc by Medvinsky and Dzierzak. In this study, the first CFU-S were first detected at the 10 dpc in the region that derives from the 9-9.5 dpc P-Sp, and now comprises the dorsal aorta (which results from the fusion of previously paired aortae), the gonads and the mesonephros (AGM). CFU-S, appeared later, after 11 dpc, in the YS and FL.[13] These data, showing the emergence of CFU-S first in the intra-embryonic region (AGM), suggested that CFU-S/HSC were generated in the embryo proper rather than in the YS, where they would secondary migrate, as suggested by previous studies performed in the avian model.[9]

Finally, these four studies coincide in that they ascribe the existence of lymphoid/multipotent precursors to an intra-embryonic site, characterized as the P-Sp/AGM region, between 8.5 and 10.5 dpc. It was later showed that before 10-11 dpc, precursors endowed with a B-lymphoid potential also harbors a erythro-myeloid and T-Lymphoid potential at the clonal level,[5] indicating that the precursors detected by Ogawa et al,[6] Cumano et al,[10] and Godin et al,[14] were multipotent.

Stem cells are precursors endowed with multipotentiality and self-renewability. This latter HSC characteristic can be assessed by the ability to reconstitute irradiated recipient mice after serial transplantation. It is also possible to measure the self-renewal capacity of a cell population in long-term reconstitution (LTR) experiments. In such assays, the investigated cell population is scored for its capacity to contribute to all hematopoietic lineages, at least 6 months after graft, and particularly the myeloid system. Indeed, myeloid cells (granulocytes and monocytes) have a short lifespan and have to be constantly generated from bone marrow LTR-HSC. This LTR potential, which is not detected in the 10 dpc AGM using conventional recipients, is then present after organ culture, meaning that LTR-HSC are autonomously generated in the AGM.[15] In the embryo, LTR activity has thus been first detected in the 10.5 dpc AGM, before it can be detected in the YS and FL at 11-11.5 dpc.[16] The presence of LTR-HSC in the YS has also been documented, at the P-Sp/AGM stages, using a different experimental set-up, namely intra-liver injection in Busulfan-treated conditioned new-born.[17,18]

These results showed that the intra-embryonic compartment contains multipotent precursors, prior to FL colonization. Furthermore, at 9 pc, these intra-embryonic multipotent precursors where shown to be restricted to the P-Sp area, since embryos devoid of both YS and P-Sp did not harbor a hematopoietic potential.[19] However, as vascular connections between the extra- and intra-embryonic compartments are well established, after 9 dpc, a cross-contamination between the two compartments could be considered.

Hematopoietic Precursors Are Present in Blood Circulation at Mid-Gestation

Moore and Owen were first to suggest, in the avian model, that hematopoietic organs have to be seeded by precursors through blood circulation.[20] The first experiment in which multipotent precursors were detected in blood circulation was achieved by Toles et al who established the existence, in 9-10 dpc circulating blood, of hematopoietic precursors capable of rescuing, in the long term (9 months), the erythroid compartment of anemic embryos injected in utero.[21] Delassus and Cumano also showed that precursors contained in blood circulation between 9.5 and 10.5 dpc were multipotent, as they are able to generate B, but also T and myeloid cells at the single cell level.[22]

In conclusion, although the first mentioned data allowed to define the P-Sp/AGM region as an intra-embryonic site containing multipotent precursors, the possibility of contamination from YS-generated precursors could not be ruled out: the precursors detected in the intra-embryonic region could result from the aggregation in the P-Sp/AGM region of YS-derived precursors brought to this site by the blood flow. Alternatively, one had to consider the possibility of a seeding of the YS by precursors independently generated in the intra-embryonic compartment.

In order to settle this question, we set out to study, before the establishment of circulation, the hematopoietic potential of the YS and the intra-embryonic splanchnopleura (Sp), which corresponds to the territory that will give rise to the P-Sp/AGM at later stages.

Establishment of Vascular Connections between the Extra- and Intraembryonic Compartment

In order to define the site of origin of intraembryonic hematopoietic precursors, which could only be ascertained by experiments carried out before the stage when blood cells can freely circulate between the extra- and intraembryonic compartments, we had to precise the stage when this event is initiated.

To this end, we marked the first circulating erythrocytes, which can be characterized by the expression the βH1 globin, specifically synthesized during early developments stages (see below). We performed wholemount in situ hybridization at stages ranging from the appearance

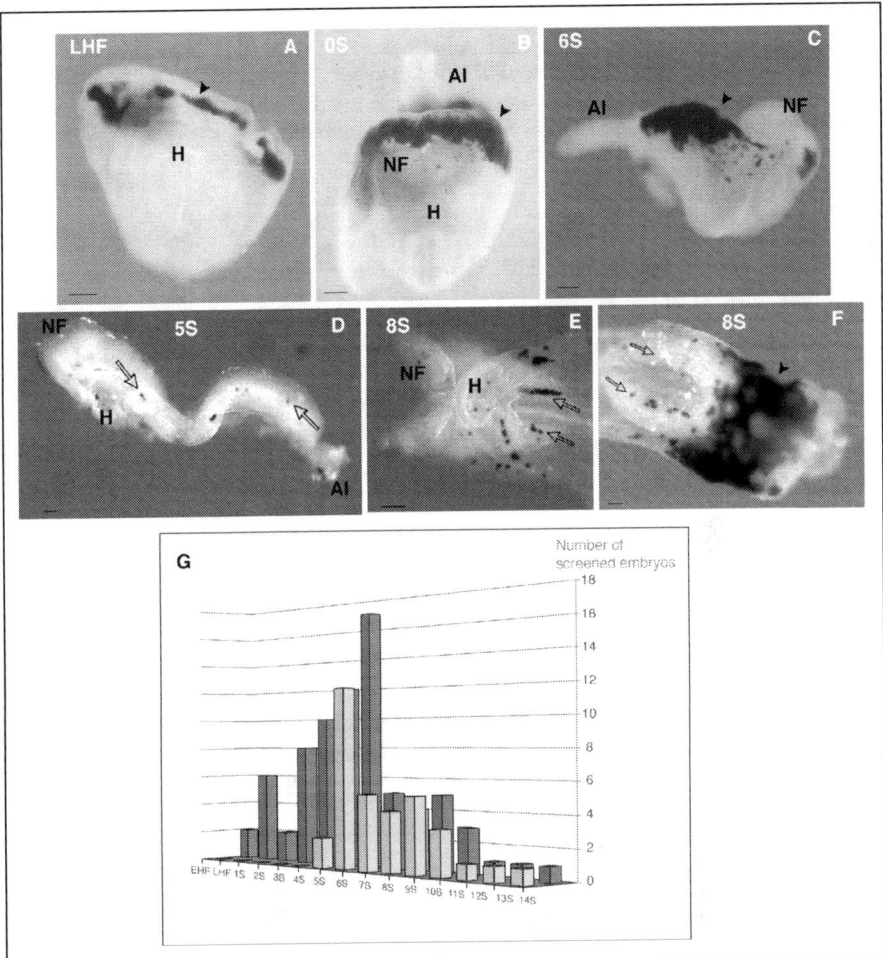

Figure 1. Establishment of blood connection between the yolk sac and the embryo. Open arrows: Circulating βH1⁺ erythrocytes; Arrowhead: βH1⁺ erythrocytes packed in YS-Blood Islands. Abbreviations: Al: Allantois; H: Heart; NF: Neural Folds. Bar = 100 μm. A-C) Progression of βH1⁺ erythrocytes from the YS-blood islands to the embryo proper: At the Late head Fold stage (LHF; A), βH1⁺cells are packed within YS-blood islands. From the 0S-stage (B) clusters of circulating cells migrate toward the intra-embryonic compartment, as blood vessel grow from the YS. This movement dramatically increases from the 6S-stage (C). D-F) Colonization of the intra-embryonic compartment by YS-derived βH1⁺ erythrocytes. In 5S embryos (D: Lateral view), circulating βH1⁺ erythrocytes are present in the aortae and in the heart. They increase in number in the aortae and heart, as large groups of βH1⁺ cells migrate from the YS-blood islands to the embryo proper. (8S-embryo, anterior view: E; caudal view: F). G) Detection of circulating βH1-expressing erythrocytes in the embryo proper at stages ranging from Early Head fold stage to 15S. In dark grey, number of analyzed embryos. In light grey, number of embryos containing βH1⁺ cells in the intra-embryonic compartment.

of βH1 erythrocytes (at 7.5 dpc), to the stage when intra-embryonic multipotent precursors are first detected (8.5 dpc: 12-15 S), and screened the embryos for the appearance of the first erythroid cells in the intra-embryonic compartment.

Table 1. *The timing of intra-embryonic colonisation by YS-derived erythrocyte is similar in embryos from both BALB/C and C57BL/6 strain. Number of embryos harbouring βH1-positive cells in the intra-embryonic compartment/ Number of tested embryos for each developmental stage.*

Stage	BALB/C	C57 BL/6	Total (%)
EHF	0/1	0/1	0/2 (0%)
LHF	0/4	0/2	0/6 (0%)
1 S	0/1	0/1	0/2 (0%)
2 S	0/3	0/5	0/8 (0%)
3 S	0/6	0/4	0/10 (0%)
4 S	1/6	1/6	2/12 (17%)
5 S	5/7	7/10	12/17 (71%)
6 S	3/3	2/2	5/5 (100%)
7 S	2/2	2/2	4/4 (100%)
8 S	1/1	1/1	2/2 (100%)
9 S		1/1	1/1 (100%)
10 S	1/1		1/1 (100%)
11 S	1/1		1/1 (100%)
14 S	1/1		1/1 (100%)

Our results, obtained by screening embryos from both Balb/c and C57BL/6, show that before the 3S-stage, no YS-derived erythroid cells are present in the intra-embryonic compartment in both strains (Fig. 1A, B, G and Table 1). The presence of βH1-expressing cells is first detected at the 4 and 5S-stages in a variable number of embryos (Fig. 1D), and the colonization of the intra-embryonic compartment is systematically achieved after the 6S-stage (Fig. 1C, E, G). This observation, previously unpublished, has since been independently confirmed.[23,24]

Origin of Intra-Embryonic Precursors: Extra- and Intraembryonic Sites Generate Independent Waves of Hematopoietic Precursors

At 8 dpc (before the 4-5 S-Stage), the Sp does not show any hematopoietic potential when cultured in vitro on a monolayer of stromal cells, after dissociation,[25] indicating that hematopoietic precursors have not yet been generated. Conversely, when early Sp are cultured in toto for a few days (3-4 days), this structure now contains hematopoietic precursors, that may be compared to those derived from similarly treated YS: we detected, by limiting dilution assays, the presence in the YS of erythro-myeloid precursors deprived of lymphoid potential, whereas the Sp proved able to provide a lympho-myeloid progeny in vitro. This data shows that the intraembryonic site is capable of autonomously generating precursors that display a lympho-myeloid potential. In contrast, the YS is only capable of providing erythro-myeloid precursors, independently of multipotent precursors.

To complete these in vitro studies of hematopoietic potential, we performed in vivo long-term reconstitution (LTR) of alymphoid Rag2$^{-/-}$γc$^{-/-}$ double knock-out recipients.[26] Using the same strategy to assess the generation of HSC during the organ culture, we observed that only the precursors derived from Sp cultured in toto were capable of LTR. In contrast, YS could only provide short-term reconstitution with a contribution restricted to the myeloid compartment: no lymphoid cells could ever be obtained from this site. Due to the absence of hematopoietic potential in the 8 dpc dissociated Sp, this data suggest that HSC are independently generated in the intraembryonic Sp (during the organ culture step), whereas the YS only supplies erythro-myeloid cells.[26]

These data allowed us to conclude that between 8.5 dpc (10-15 S-Stage) and 10.5 dpc, multipotent precursors (HSC) are generated in the P-Sp/AGM region, released in the aortic blood circulation, from where they seed the FL, but also the YS. These multipotent hematopoietic cells, detected as soon as 9-9.5 dpc in the YS, are capable of LTR when injected in the liver of busulfan-treated newborn recipients, as well as those collected from the P-Sp,[18] (See also Yoder contribution to the present issue). Thus, the presence of HSC in the YS appears to result from a contamination from intraembryonic region through blood circulation.

What's Left to Do for the YS?

We have previously seen that the YS is the first hematopoietic site during mouse development, as soon as 7 dpc. Thus, as HSC production only starts at 8.5-9 dpc in the intraembryonic compartment, extra-embryonic YS hematopoiesis appears to be HSC-independent, which rises many questions about the origin of these cells as compared to adult definitive hematopoiesis, as it was defined in the fetal liver and adult bone marrow.

"Definitive" hematopoiesis consists of the diverse processes at the origin of the various hematopoietic lineages in the adult. In contrast, "primitive" hematopoiesis only generates hematopoietic subsets that diverge from their adult "definitive" counterparts by morphological and physiological aspects.

The first hematopoietic cells detected in the YS consist of primitive nucleated erythrocytes, which are generated in the blood islands. These primitive red blood cells express a different set of hemoglobins, compared to adult erythrocytes, the so-called fetal hemoglobins (βH1, ζ and ϵ) that display a higher affinity for dioxygen, then allowing the embryo to absorb dioxygen from maternal blood. When blood circulation is established, these erythrocytes can circulate while remaining nucleated. Recent studies have shown that they enucleate in blood circulation, a few days later, at 13-14 dpc.[27] After the onset of FL hematopoiesis, at 10-11 dpc, definitive erythrocytes are now produced, that enucleate before release into the circulation, and start to express adult-type hemoglobins (β-major and α-globins).

Naito and Takahashi also used the terms primitive and definitive to discriminate two waves of YS macrophages. The first one, that appears early after gastrulation and rapidly gives rise to a differentiated progeny, seems to by-pass the pro-monocytes and monocytes stages described in the adult bone marrow, and was thus qualified as "primitive" or fetal. In contrast, the second lineage was shown to follow the same precursors succession as identified in the bone marrow.[28,29] Finally, it has been shown that early macrophages in the mouse YS express low levels of lysozyme M, as compared to their adult counterparts.[30]

A third subset of "primitive" hematopoietic cells has been identified in the YS, within the megakaryocytic lineage. The primitive megakaryocytes differentiate very fast from YS precursors and display smaller size and ploidy than adult ones.[31]

Primitive hematopoiesis thus produces terminally differentiated cells that acquire different characteristics, regarding to their adult "definitive" counterparts. Molecular requirements for their differentiation also seems independent from the ones leading to the establishment of definitive hematopoiesis. Indeed, the invalidation of Runx1/AML-1 or c-Myb, amongst others (for review see ref. 32), only affects the generation of definitive HSC in the embryo proper, whereas YS hematopoiesis is not affected.[33-37] It was also shown that YS "primitive" macrophage development is not affected by the absence of PU.1, which is required for the differentiation of "definitive" macrophages.[38]

The generation of "definitive" erythro-myeloid precursors has also been documented in the YS (see Chapter by Palis). Clonogenic myeloid precursors can differentiate into definitive adult-type erythrocytes, mast cells, megakaryocytes, granulocytes and macrophages. These YS-derived precursors are the first to seed the FL as early as 9.5-10 dpc.[39,40] and have the potential to differentiate in this microenvironment (unpublished data from JYB). This YS contribution to FL hematopoiesis may correspond to a transient way to produce blood cells between the YS primitive production and the red blood cells production from HSC, that colonize the FL at later stages.

Another important contribution from the YS appears to be the production of microglia in the nervous system. In the avian system, it was indeed shown that YS-derived macrophages are able to seed the brain and retina.[41,42] In the adult mouse, microglia is a stable, quiescent system,[43] not renewed from bone marrow precursors in steady state conditions.[44] Alliot et al have shown the existence, in the early YS, of hematopoietic precursors that can differentiate into microglia when cultured onto a layer of astrocytes. The kinetic of appearance of YS-derived microglia precursors, and brain microglia suggested that microglia could be entirely generated from YS precursors.[45] While testing the capacity of YS-derived macrophages to contribute to adult microglia, we further showed that YS macrophages can engraft for more than 6 months in the brain of newborn recipients either injected intra-cerebrally or intravenously (unpublished observations from JYB). Thus YS is able to contribute to the establishment of adult microglia.

Intraembryonic Hematopoiesis: To the Precise Origin of HSC

In this section, we will focus on the correlation between the structure of the AGM region in the mammal embryos, the phenotype and differentiation potential of hematopoietic precursors present in this site, the genetic characterization of these precursors, and the in situ pattern of expression obtained for gene involved in HSC generation. This strategy aims at defining a model for de novo production of HSC in the P-Sp/AGM region.

We previously established that the P-Sp/AGM region was producing multipotent hematopoietic precursors during a short time window, between 8.5 and 12 dpc. Through limiting-dilution assays, we quantified the number of multipotent cells in this site, and established that AGM production reaches a peak at 10.5-11 dpc. At this stage, the AGM contains around 100-150 multipotent cells, all of them within the AA4.1+ fraction.[19] This kinetic allowed us to evaluate the number of clonogenic multipotent cells produced de novo at about 500 per embryo. This number, although initially low, will increase during the expansion phase that takes place at later stages in the FL (for a review, see ref. 46). A possible increase of HSC pool size, through mitotic activity (versus de novo generation) within the AGM has not been excluded since multipotent cells are active in cell cycle at 10-11 dpc (unpublished observation from JYB). Nevertheless, the P-Sp/AGM is not a site of hematopoietic differentiation, as previously shown.[25]

Spatial Organization of the AGM in Mammals

As mentioned above, the 10 dpc AGM region, that contains the aorta, the genital ridges and the mesonephros, derives from the P-Sp region and is organized around the dorsal aorta and its ramifications, such as the omphalo-mesenteric and umbilical arteries (Fig. 2A). This region has been identified as a hemogenic site in all vertebrate embryos, and is now extensively studied from the zebrafish to the human embryos (see corresponding sections in this issue).

Both in vitro[5] and in vivo[47] demonstrated that AGM hematopoietic precursors were concentrated in the aorta and underlying splanchnic mesoderm rather than in the uro-genital compartment. This data well correlates with those obtained in the avian model, where the ventral region of the aorta was shown to directly derive from the Sp, in contrast to the dorso-lateral

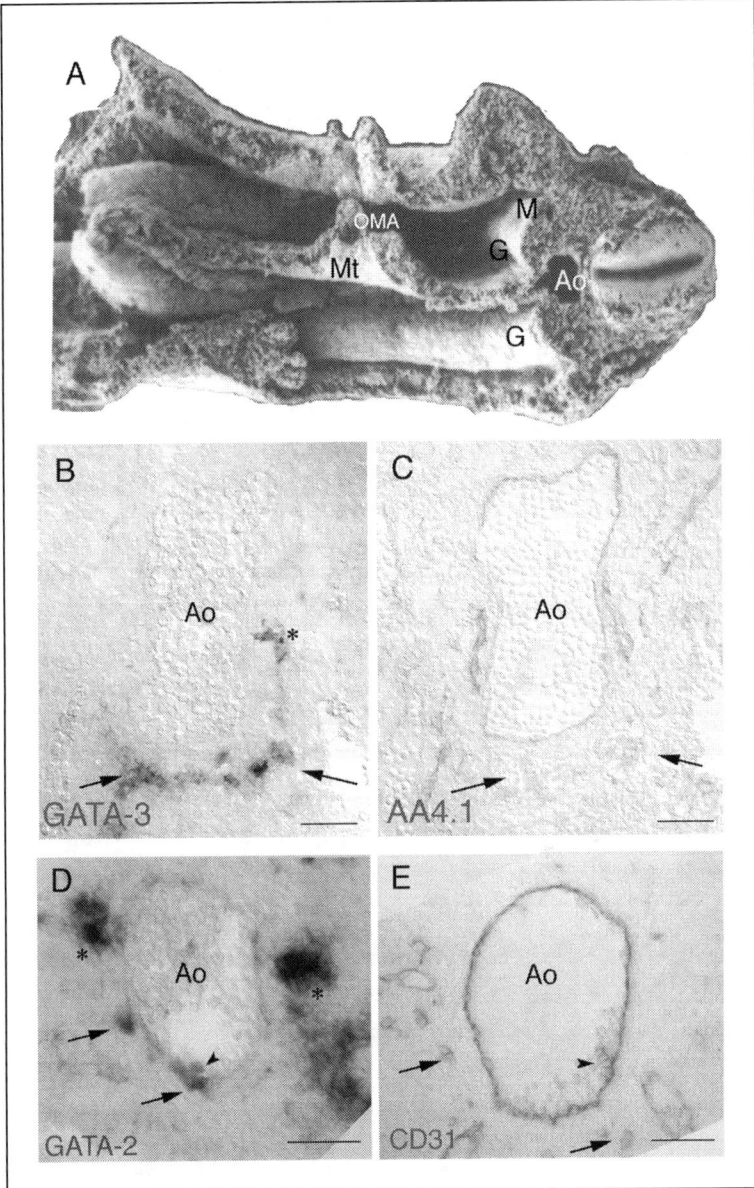

Figure 2. HSC localization within 10-10.5 dpc AGM. Arrow: Sub-aortic patches; Arrowhead: Hematopoietic intra-aortic clusters. Abbreviations: Ao: Aorta; G: Gonads; M: Mesonephros; Mt: Mesentery; OMA: Omphalo-mesenteric artery. A) Scanning Electron Microscopy picture displaying the structure of the AGM. The 10-10.5 dpc sub-aortic patches (arrows) underlying the aorta express GATA-3 (B) and GATA-2 (D) mRNA, as well as the AA4.1 (C) and CD31 (E) antigens. B and C are consecutive sections of the same embryo. GATA-2 (D), AA4.1 (C) and CD31 (E) also label endothelial cells. Sympathetic ganglia (Asterisk) located laterally to the aorta also express GATA-3 (B) and GATA-2 (D) mRNA. Hematopoietic intra-aortic clusters also express these markers, as shown here for GATA-2 (D) and CD31 (E). Bar= 50 µm.

part that is of somitic origin.[48] Using the quail-chick chimaera experimental set-up, it was also demonstrated that only the Sp-derived cells has the potential to contribute to hematopoiesis.[48]

At the beginning of the 20th century, Emmel pointed out the presence, in chick embryos, of cell clusters located on the floor of the aorta and identified these cells as hematopoietic.[49] These aortic clusters, shown to be present in every studied vertebrate species, have now been characterized in terms of cell surface markers expression: the cell in the clusters are hematopoietic, since they express the pan-leukocyte marker CD45, although at low levels,[50-52] and the adhesion molecule CD34 and CD31 (Fig. 2E),[53,54] among others,[55] (for a review, see ref. 32). These clusters appear transiently in the AGM between 10 and 12 dpc,[52] which correspond the peak of multipotent hematopoietic precursor production[5] (Fig. 3). At the level of intra-aortic clusters, the endothelium and the basal membrane are interrupted, as shown by electron microscopy performed on this region in mouse, chick and human embryos.[51,56,57]

Until recently, the intra-aortic clusters were the only structures supposedly involved in HSC generation. We recently identified in the mouse P-Sp/AGM region, another structure potentially involved in this process (Figs. 2B-E and 3). Through immuno-histological and in situ hybridization studies, we identified the existence of sub-aortic patches[52] (For a review, see ref. 32), by the expression of GATA-3 (Fig. 2), GATA-2 (Fig. 2), two transcription factors expressed by hematopoietic precursors, and of AA4.1 (Fig. 2), an adhesion molecule expressed on early hematopoietic precursors.[2,19,58,59] Such structures were also identified in the human embryo at the time of multipotent hematopoietic precursor production[54] (see also the contributionby Marshall to this issue). Various observations argue for a participation of these sub-aortic patches to the process of intraembryonic-HSC generation (Fig. 3): 1 the timing of sub-aortic patches presence in the P-Sp/AGM strictly correlates with its hemogenic activity, since they appear at 8.5 dpc with the generation of the first multipotent precursor[19] and disappear with the cessation of this activity at 12 dpc;[5] 2 The sub-aortic patches are only present in the splanchnic mesoderm underlying the aortic floor, mostly below the aortic hematopoietic cell clusters previously described.

In Vitro and in Vivo Characterization of Intraembryonic Hematopoietic Stem Cells

As for the adult HSC, which phenotype has been elucidated in the bone marrow, it was interesting to purify HSC at their site of de novo production, the AGM region. The AGM phenotypic characterization was mainly performed in the mouse embryo, due to the possibility to purify cell subsets using monoclonal antibodies. The obtained fractions were then tested either for their capacity to generate lymphoid progenies in vitro (through limiting dilution or single cell assays), or for their potential to reconstitute the lympho-myeloid system of irradiated recipients (LTR experiments).

The first fractionation, leading to a low level of HSC enrichment, allowed to establish that the hematopoietic potential of the 10-11 dpc AGM was contained in the cKit[+] CD34[+] fraction.[22,60] We showed that the multipotent hematopoietic precursors were enriched at a ratio of 1:4 in the AA4.1[+] fraction of the 10 dpc AGM.[5]

Recently, we obtained a higher level of HSC purification, by using a combination of five different markers. We show that in the CD45[-/lo] c-Kit[+] AA4.1[+] CD31[+] CD41[+] subset, up to 1:1.4 cells is a HSC (characterized in vitro as multipotent at the single cell level and in vivo, as capable of LTR when injected into irradiated NK-deficient Rag2[-/-]γc[-/-] recipients), whereas only 1:7 cells is restricted to the erythro-myeloid lineage.[55] In terms of absolute numbers, this enrichment equals the previously quantified number of AGM-multipotent precursors (100-150) at 10 dpc,[5] implying that this subset accounts for the bulk of HSC present at this stage.

This high level of purification allowed us to perform gene expression analysis. We showed that HSC in this subset coexpress the transcription factors GATA-2, GATA-3 and Lmo-2.[55]

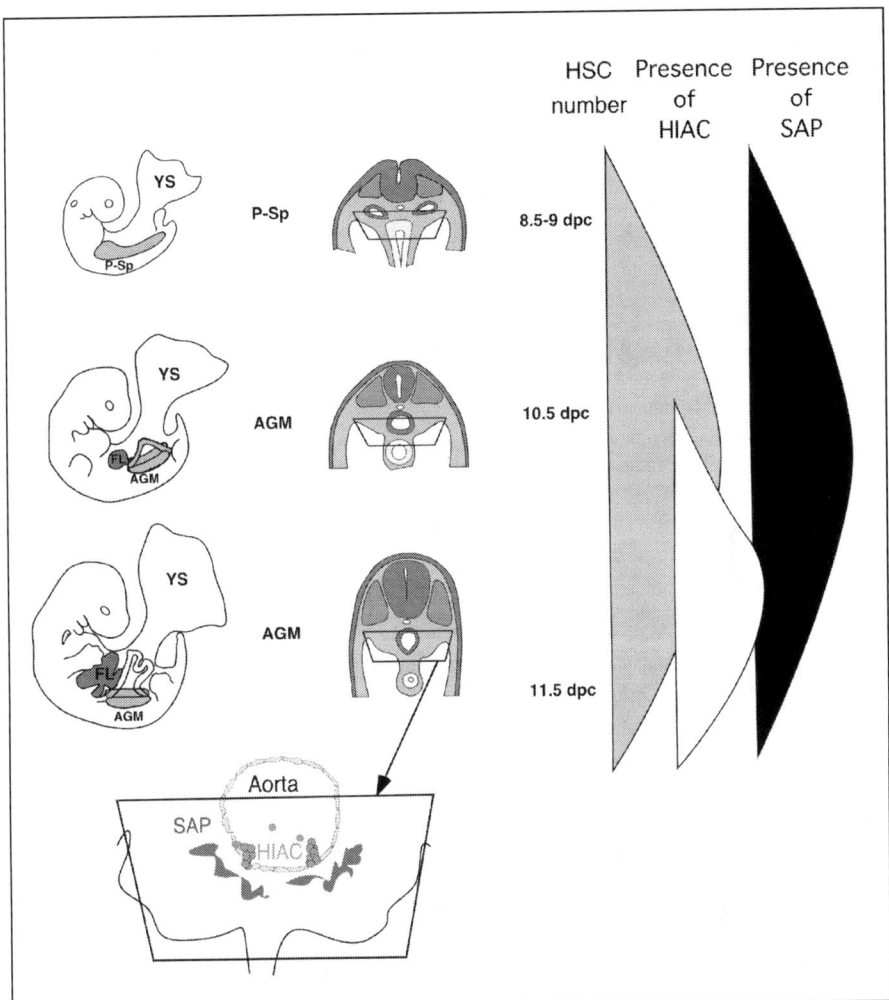

Figure 3. Localization of the intra-embryonic hemogenic site. Abbreviations: SAP: Sub-aortic patches; HIAC: Hematopoietic intra-aortic clusters. In the left column, the location within the whole embryo of the P-Sp/AGM is shown at different developmental stages. The middle column displays the P-Sp/AGM anatomy on schematic sections at the corresponding stage. In the right column summarizes: 1 the evolution of HSC number between 8.5 and 12 dpc (grey) with a peak at 10.5-11dpc; 2 the relative presence in the embryo of HIAC and SAP during the phase of HSC generation. Whereas the SAP (black) are present during the whole phase of HSC generation, HIAC (white) are only detected at the peak of this production.

Similar results were previously obtained when CD45^{lo} CD34$^+$ cells, isolated from the human AGM, were characterized for gene expression, as this subset was shown to express transcripts for c-Myb, Tal-1/SCL, GATA-2 and GATA-3.[61] In the mouse embryo, it was shown at the clonal level, that c-Kit$^+$ CD34$^+$ isolated from the AGM region coexpress Lmo-2, AML-1/Runx-1, PU-1 and GATA-2.[22]

From studies performed in the mouse and human models, it can be concluded that HSC are highly enriched in the CD45$^{-/lo}$ c-Kit$^+$ CD34$^+$ CD31$^+$ AA4.1$^+$ CD41$^+$, and express several transcription factors such as GATA-2, GATA-3, Tal-1, Runx-1 and Lmo-2.

In Situ Localization of HSC in the Mouse AGM Region

The correlation of data obtained by our group and others allowed to define a specific expression pattern for HSC purified at their site of origin, the AGM. Among surface markers and transcription factors expressed by AGM-HSC, only CD41 and GATA-3 appear to specifically individualize HSC from endothelial cells in this region. GATA-3 expression pattern, already analyzed in the AGM region through in situ hybridization,[52] contributed to the identification of the previously mentioned sub-aortic patches in the splanchnic mesoderm underlying the aorta. CD41 expressing cells were previously identified in intra-aortic clusters in the chick embryo.[62] We recently allocated CD41$^+$ CD31$^+$ HSC, through confocal immuno-histological studies, to intra-aortic clusters, which actually seemed heterogeneous, since CD31$^+$ CD41$^-$ could also be identified in these structures.[55] Moreover, we could show the presence within the ventral GATA-3$^+$ CD31$^+$ CD41$^-$ sub-aortic patches,[52,55] of individual CD41$^+$ cells, that appeared distributed along a continuum between sub-aortic patches and intra-aortic clusters upon tri-dimensional reconstruction of confocal plans.

Thus, in accordance with data published in the chick and mammals,[51,56,57] we confirm the presence of HSC (multipotent hematopoietic precursors) within the intra-aortic clusters, and further show that HSC are also present within the sub-aortic patches. We described the sub-aortic patches in terms of surface markers and transcription factors expression: thus, these nonvascular structures[55] can be identified as CD45$^-$ CD41$^-$ CD31$^+$ AA4.1$^+$ GATA-2$^+$ GATA-3$^+$ in the mouse embryo.[52,55] The function played by these structures during the HSC generation process remains to be analyzed.

An Alternative Model to the Hemogenic Endothelium

Until now, the mechanisms leading to the de novo HSC production in the AGM have not been precisely elucidated. Many published studies are indicative of a developmental relationship between endothelial and hematopoietic cells, based on common surface markers and transcription factors shared by these two cell types (for a review, see ref. 32), both in the YS and in the AGM. Data obtained from ES cell studies point to a common precursor for endothelial and hemato-poietic cells, the hemangioblast,[63] which is now considered to be present in the YS-blood islands and to give rise to the extra-embryonic endothelial and hematopoietic cells. HSC generation is closely linked to the aortic floor, at a time when the vascular network is differentiated, thus disqualifying a putative hemangioblasts as the source of HSC. The possibility that endothelial cells from the aortic floor may transdifferentiate to give rise to HSC has thus been considered, in chick, mouse and human embryos[51,64-68] (see chapter by Jaffredo and Dzierzak). Most studies aiming at testing this "hemogenic endothelium" model are based on the purification of putative endothelial cells, CD45- CD34$^+$ CD31$^+$ (characterized as such by the lack of expression of the "pan-leukocyte" marker CD45). This subset has been shown capable of hematopoietic differentiation when cultured in vitro.[64-66] Since, we established that such a phenotype could be independent of an endothelial phenotype as HSC in the AGM also are CD45$^-$ CD31$^+$ CD34$^+$ and additionally express CD41,[55] the latter marker being strictly ascribed to the hematopoietic lineage, appearing before CD45 during development.[69]

In another experimental set-up, in vivo lineage analyses were performed by injecting DiI-labeled acetylated-Low Density Lipoprotein (Ac-LDL), in early chick[51] and mouse[68] embryos at stages when intra-aortic clusters are hardly detectable. In these studies, it was postulated that only endothelial cells and, to lesser extent, macrophages, could up-take Ac-LDL. One day after Ac-LDL injection, DiI-labelled intra-aortic cluster could be observed in the chick embryo[51] and circulating hematopoietic cells (erythrocytes) could be detected in the mouse embryo.[68] Based on the assumption that only endothelial cells can uptake Ac-LDL, the authors concluded that the clusters of hematopoietic precursors originate from the ventral endothelium itself. Since, we have shown that purified HSC (c-Kit$^+$ CD41$^+$) also could incorporate Ac-LDL.[55] It thus appears that, at least during the embryonic stages considered here, Ac-LDL uptake is not a feature specific to endothelial cells and macrophages, as initially thought.

Due to these observations and the fact that HSC could be localized in other structures than the intra-aortic clusters, the sub-aortic patches, we propose an alternative model (Fig. 3) to the "hemogenic endothelium" model.

We previously quantified the number of multipotent precursors in the P-Sp/AGM region, through limiting-dilution assays.[5] This quantification showed that the first multipotent precursors were produced in the P-Sp from the 15 S-stage onwards, thus long before the intra-aortic clusters can be detected in the dorsal aorta. In contrast, the sub-aortic patches are already present by that stage, and persist during the whole period of HSC production in the AGM, from 8.5 to 12 dpc.[52] This prompted us to propose the sub-aortic patches as the site of HSC production in the AGM region.[55] As suggested in the human embryo,[54,70] the sub-aortic patches could form a suitable environment for HSC emergence, presumably from a mesodermic precursor still unidentified. Then, HSC would migrate to the aortic lumen, where they might form the intra-aortic clusters; this migration could explain the CD41 continuum observed between the sub-aortic patches and the aortic clusters,[55] from where they would enter blood circulation to colonize hematopoietic organs, such as the fetal liver and the thymus (Fig. 4).

Conclusion

The YS is the first site of hematopoietic production during embryogenesis. For this reason, it was long considered as the emergence site for adult "definitive" hematopoietic stem cells. We have shown here that the YS and the intraembryonic splanchnopleura could give rise to two independent waves of hematopoietic precursors. On one hand, the YS produces distinct types of primitive and definitive erythro-myeloid-restricted precursors.[39,40] On the other hand, the Sp region produces HSC, endowed with LTR capacity.[26] Thus, YS hematopoiesis occurs in the absence of multipotent hematopoietic precursors.

HSC production starts at the 10-15 somite-stage, in the intraembryonic P-Sp. This transient process is achieved by 12 dpc. We have characterized the newly produced HSC at their site of de novo production.[55] The levels of enrichment obtained (almost 1:1) allowed to ascribe a particular gene expression pattern to this subset, which permitted to localize these cells in the AGM region. Thus, HSC are not only detectable in the previously described intra-aortic clusters, but also in sub-aortic patches underlying the aortic floor. These structures are present during the whole time of HSC generation.[52] This prompted us to propose that these structures are involved in the process of HSC de novo production, from unidentified precursors (Fig. 3).

In the human embryo, a similar dichotomy between the YS and the intraembryonic Sp hematopoietic generation was also observed.[71] Although the existence of structures similar to the sub-aortic patches was also reported,[54,70] the involvement of these structures in the emergence of HSC was no further documented.

Primitive hematopoietic cells produced in the YS are thought to rapidly provide the developing embryo with blood cells. This contribution is limited compared to FL hematopoiesis. Palis and collaborators[72] have shown that clonogenic "definitive" erythro-myeloid precursors arising in the YS could be detected in the FL as soon 10 dpc, concomitantly to FL colonization by AGM-HSC. We previously determined that the P-Sp/AGM region produced a pool of about 500-700 HSC.[5] This limited subset is thought to undergo many rounds of symetric cell divisions in the fetal liver to expand the pool of hematopoietic multipotent precursors (for a review see ref. 46). We propose that YS erythro-myeloid precursor can account for FL hematopoiesis during an early HSC expansion phase, before the HSC pool is ready to supply the organism with differentiated cells. From the FL, HSC are known to colonize other hematopoietic

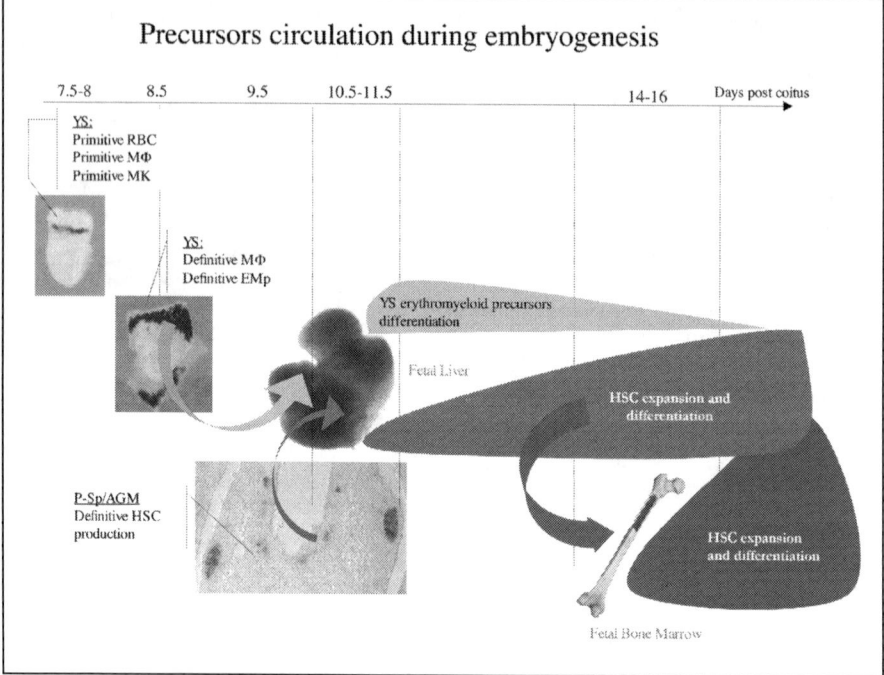

Precursors circulation during embryogenesis

Figure 4. Hematopoietic precursor circulation during embryogenesis. The YS produces primitive-type hematopoietic cells as soon as 7-7.5 dpc. The production of definitive macrophages and erythromyeloid precursors occurs at later stages (8 dpc), independently of HSC. At 10 dpc, these YS-derived precursors colonize the FL, where they differentiate into all myeloid lineages. From 8.5-9 dpc, the P-Sp/AGM produces definitive adult-type HSC, which migrate to the FL, where they expand and differentiate, before they colonize the fetal bone marrow, where they reside for the whole post-natal life.

organs, such has the fetal spleen[4,5] and the fetal bone marrow, from 15-16 dpc onwards.[6] After birth, the bone marrow will remain the only source of hematopoietic cells for the whole life of the animal (Fig. 4).

Acknowledgements

This work was supported by grants from Fondation pour la Recherche Médicale (n° INE200330307040, Ligue Régionale contre le Cancer Comité des Yvelines (n°GL/CB/0167-02) and from Institut Gustave Roussy (CRI-SPS-2003-02) to I. G.

References

1. Gothert JR, Gustin SE, Hall MA et al. In vivo fate-tracing studies using the SCL stem cell enhancer: Embryonic hematopoietic stem cells significantly contribute to adult hematopoiesis. Blood 2005; 105(7):2724-2732.
2. Jordan CT, McKearn JP, Lemischka IR. Cellular and developmental properties of fetal hematopoietic stem cells. Cell 1990; 61(6):953-63.
3. Metcalf D, Moore MAS. Embryonic aspects of haemopoiesis. In: Neuberger A, Tatum EL, eds. Haematopoietic cells. Amsterdam, London: North Holland Publish Co, 1971:173-271.
4. Christensen JL, Wright DE, Wagers AJ et al. Circulation and chemotaxis of fetal hematopoietic stem cells. PLoS Biology 2004; 2(3):368-377.

5. Godin I, Garcia-Porrero JA, Dieterlen-Lievre F et al. Stem cell emergence and hemopoietic activity are incompatible in mouse intraembryonic sites. J Exp Med 1999; 190(1):43-52.

6. Ogawa M, Nishikawa S, Yamamura F et al. B cell ontogeny in murine embryo studied by a culture system with the monolayer of a stromal cell clone, ST2: B cell progenitor develops first in the embryonal body rather than in the Yolk sac. EMBO 1988; 7:1337-1343.

7. Johnson GR, Moore M. Role of stem cell migration in initiation of mouse foetal liver haemopoiesis. Nature 1975; 258:726-729.

8. Houssaint E. Differentiation of the mouse hepatic primordium: II- Extrinsic origin of the haemopoitic cell line. Cell Diff 1981; 10:243-252.

9. Dieterlen-Lièvre F. On the origin of haemopoietic stem cells in the avian embryo: An experimental approach. J Embryol Exp Morphol 1975; 33:607-619.

10. Cumano A, Furlonger C, Paige C. Differentiation and characterisation of B-cell precursors detected in the yolk sac and embryo body of embryos beginning at the 10-12 somite stage. Proc Nat Acad Sci USA 1993; 90:6429-6433.

11. Godin IE, Garcia-Porrero JA, Coutinho A et al. Para-aortic splanchnopleura from early mouse embryos contains B1a cell progenitors. Nature 1993; 364(6432):67-70.

12. Till JE, Mc CE. A direct measurement of the radiation sensitivity of normal mouse bone marrow cells. Radiat Res 1961; 14:213-22.

13. Medvinsky AL, Samoylina NL, Muller AM et al. An early preliver intraembryonic source of CFU-S in the developing mouse. Nature 1993; 364(6432):64-7.

14. Godin I, Garcia Porrero JA, Coutinho A et al. Paraaortic splanchnopleura contains B1a lymphocyte precursors. Nature 1993; 364:67-69.

15. Medvinsky AL, Gan OI, Semenova ML et al. Development of day-8 colony-forming unit-spleen hematopoietic progenitors during early murine embryogenesis: Spatial and temporal mapping. Blood 1996; 87(2):557-66.

16. Muller AM, Medvinsky A, Strouboulis J et al. Development of hematopoietic stem cell activity in the mouse embryo. Immunity 1994; 1(4):291-301.

17. Yoder MC, Hiatt K, Dutt P et al. Characterization of definitive lymphohematopoietic stem cells in the day 9 murine yolk sac. Immunity 1997; 7:335-344.

18. Yoder MC, Hiatt K, Mukherjee P. In vivo repopulating hematopoietic stem cells are present in the murine yolk sac at day 9.0 postcoitus. Proc Natl Acad Sci USA 1997; 94:6776-6780.

19. Godin I, Dieterlen-Lièvre F, Cumano A. Emergence of multipotent hematopoietic cells in the yolk sac and paraaortic splanchnopleura in mouse embryo, beginning at 8.5 days postcoitus. Proc Natl Acad Sci USA 1995; 92(3):773-777.

20. Moore M, Owen J. Chromosome marker studies on the development of the haemopoietic system in the chick embryo. Nature 1965; 208:956-989.

21. Toles JF, Chui DH, Belbeck LW et al. Hemopoietic stem cells in murine embryonic yolk sac and peripheral blood. Proc Natl Acad Sci USA 1989; 86(19):7456-9.

22. Delassus S, Cumano A. Circulation of hematopoietic progenitors in the mouse embryo. Immunity 1996; 4:97-106.

23. Downs KM, Gifford S, Blahnik M et al. Vascularization in the murine allantois occurs by vasculogenesis without accompanying erythropoiesis. Development 1998; 125(22):4507-20.

24. McGrath KE, Koniski AD, Malik J et al. Circulation is established in a stepwise pattern in the mammalian embryo. Blood 2003; 101(5):1669-76.

25. Cumano A, Dieterlen-Lievre F, Godin I. Lymphoid potential, probed before circulation in mouse, is restricted to caudal intraembryonic splanchnopleura. Cell 1996; 86(6):907-16.

26. Cumano A, Ferraz JC, Klaine M et al. Intraembryonic, but not yolk sac hematopoietic precursors, isolated before circulation, provide long-term multilineage reconstitution. Immunity 2001; 15(3):477-85.

27. Kingsley PD, Malik J, Fantauzzo KA et al. Yolk sac-derived primitive erythroblasts enucleate during mammalian embryogenesis. Blood 2004; 104(1):19-25.

28. Naito M, Yamamura F, Nishikawa S et al. Development, differentiation, and maturation of fetal mouse yolk sac macrophages in cultures. J Leukoc Biol 1989; 46(1):1-10.

29. Takahashi, Yamamura, Naito. Differentiation, maturation, and proliferation of macrophages in the mouse yolk sac: A light-microscopic, enzyme-cytochemical, immunohistochemical, and ultrastructural study. J Leukocyte Biol 1989; 45:87-96.
30. Faust N, Huber MC, Sippel AE et al. Different macrophage populations develop from embryonic/fetal and adult hematopoietic tissues. Exp Hematol 1997; 25(5):432-44.
31. Xu MJ, Matsuoka S, Yang FC et al. Evidence for the presence of murine primitive megakaryocytopoiesis in the early yolk sac. Blood 2001; 97(7):2016-22.
32. Godin I, Cumano A. The hare and the tortoise: An embryonic haematopoietic race. Nat Rev Immunol 2002; 2(8):593-604.
33. Castilla LH, Wijmenga C, Wang Q et al. Failure of embryonic hematopoiesis and lethal hemorrhages in mouse embryos heterozygous for a knocked-in leukemia gene CBFB-MYH11. Cell 1996; 87(4):687-96.
34. Wang Q, Stacy T, Binder M et al. Disruption of the Cbfa2 gene causes necrosis and hemorrhaging in the central nervous system and blocks definitive hematopoiesis. Proc Natl Acad Sci USA 1996; 93(8):3444-9.
35. Yergeau DA, Hetherington CJ, Wang Q et al. Embryonic lethality and impairment of haematopoiesis in mice heterozygous for an AML1-ETO fusion gene. Nat Genet 1997; 15(3):303-6.
36. North T, Gu TL, Stacy T et al. Cbfa2 is required for the formation of intra-aortic hematopoietic clusters. Development 1999; 126(11):2563-75.
37. Mucenski ML, McLain K, Kier AB et al. A functional c-myb gene is required for normal murine fetal hepatic hematopoiesis. Cell 1991; 65(4):677-89.
38. Lichanska AM, Browne CM, Henkel GW et al. Differentiation of the mononuclear phagocyte system during mouse embryogenesis: The role of transcription factor PU.1. Blood 1999; 94(1):127-38.
39. Palis J, Robertson S, Kennedy M et al. Development of erythroid and myeloid progenitors in the yolk sac and embryo proper of the mouse. Development 1999; 126(22):5073-84.
40. Palis J, Chan RJ, Koniski A et al. Spatial and temporal emergence of high proliferative potential hematopoietic precursors during murine embryogenesis. Proc Natl Acad Sci USA 2001; 98(8):4528-33.
41. Cuadros M, A M, G MP et al. Microglia in the mature and developing quail brain as revealed by a monoclonal. Neurosci Lett 1992; 148:11-14.
42. Cuadros MA, Moujahid A, Quesada A. Development of microglia in the quail optic tectum. J Comp Neurol 1994; 346:1.
43. Lawson LJ, Perry VH, Gordon S. Turnover of resident microglia in the normal adult mouse brain. Neuroscience 1992; 48(2):405-15.
44. Kennedy DW, Abkowitz JL. Kinetics of central nervous system microglial and macrophage engraftment: Analysis using a transgenic bone marrow transplantation model. Blood 1997; 90(3):986-93.
45. Alliot F, Godin I, Pessac B. Microglia derive from progenitors, originating from the yolk sac, and which proliferate in the brain. Brain Res Dev Brain Res 1999; 117(2):145-52.
46. Lessard J, Faubert A, Sauvageau G. Genetic programs regulating HSC specification, maintenance and expansion. Oncogene 2004; 23(43):7199-209.
47. de Bruijn MF, Speck NA, Peeters MC et al. Definitive hematopoietic stem cells first develop within the major arterial regions of the mouse embryo. EMBO J 2000; 19(11):2465-74.
48. Pardanaud L, Luton D, Prigent M et al. Two distinct endothelial lineages in ontogeny, one of them related to hemopoiesis. Development 1996; 122(5):1363-71.
49. Emmel V. The cell clusters in the dorsal aorta of mammalian embryos. Am J Anat 1916; 401-421.
50. Tavian M, Coulombel L, Luton D et al. Aorta-associated CD34+ hematopoietic cells in the early human embryo. Blood 1996; 87(1):67-72.
51. Jaffredo T, Gautier R, Eichmann A et al. Intraaortic hemopoietic cells are derived from endothelial cells during ontogeny. Development 1998; 125(22):4575-83.
52. Manaia A, Lemarchandel V, Klaine M et al. Lmo2 and GATA-3 associated expression in intraembryonic hemogenic sites. Development 2000; 127(3):643-53.
53. Garcia-Porrero JA, Manaia A, Jimeno J et al. Antigenic profiles of endothelial and hemopoietic lineages in murine intraembryonic hemogenic sites. Dev Comp Immunol 1998; 22(3):303-19.

54. Marshall CJ, Moore RL, Thorogood P et al. Detailed characterization of the human aorta-gonad-mesonephros region reveals morphological polarity resembling a hematopoietic stromal layer. Dev Dyn 1999; 215(2):139-47.

55. Bertrand JY, Giroux S, Golub R et al. Characterization of purified intraembryonic hematopoietic stem cells as a tool to define their site of origin. PNAS 2005; 102(1):134-139.

56. Garcia-Porrero JA, Godin IE, Dieterlen-Lievre F. Potential intraembryonic hemogenic sites at preliver stages in the mouse. Anat Embryol (Berl) 1995; 192(5):425-35.

57. Tavian M, Cortes F, Charbord P et al. Emergence of the haematopoietic system in the human embryo and foetus. Haematologica 1999; 84(Suppl EHA-4):1-3.

58. Cumano A, Paige CJ. Enrichment and characterization of uncommitted B-cell precursors from fetal liver at day 12 of gestation. EMBO J 1992; 11(2):593-601.

59. Petrenko O, Beavis A, Klaine M et al. The molecular characterization of the fetal stem cell marker AA4. Immunity 1999; 10:691-700.

60. Sanchez MJ, Holmes A, Miles C et al. Characterization of the first definitive hematopoietic stem cells in the AGM and liver of the mouse embryo. Immunity 1996; 5(6):513-25.

61. Labastie MC, Cortes F, Romeo PH et al. Molecular identity of hematopoietic precursor cells emerging in the human embryo. Blood 1998; 92(10):3624-35.

62. Ody C, Vaigot P, Quere P et al. Glycoprotein IIb-IIIa is expressed on avian multilineage hematopoietic progenitor cells. Blood 1999; 93(9):2898-906.

63. Choi K. Hemangioblast development and regulation. Biochem Cell Biol 1998; 76(6):947-56.

64. Nishikawa SI, Nishikawa S, Kawamoto H et al. In vitro generation of lymphohematopoietic cells from endothelial cells purified from murine embryos. Immunity 1998; 8(6):761-9.

65. Oberlin E, Tavian M, Blazsek I et al. Blood-forming potential of vascular endothelium in the human embryo. Development 2002; 129(17):4147-57.

66. North TE, de Bruijn MF, Stacy T et al. Runx1 expression marks long-term repopulating hematopoietic stem cells in the midgestation mouse embryo. Immunity 2002; 16(5):661-72.

67. de Bruijn MF, Ma X, Robin C et al. Hematopoietic stem cells localize to the endothelial cell layer in the midgestation mouse aorta. Immunity 2002; 16(5):673-83.

68. Sugiyama D, Ogawa M, Hirose I et al. Erythropoiesis from acetyl LDL incorporating endothelial cells at the preliver stage. Blood 2003; 101(12):4733-8.

69. Mikkola HK, Fujiwara Y, Schlaeger TM et al. Expression of CD41 marks the initiation of definitive hematopoiesis in the mouse embryo. Blood 2003; 101(2):508-16.

70. Marshall CJ, Kinnon C, Thrasher AJ. Polarized expression of bone morphogenetic protein-4 in the human aorta-gonad-mesonephros region. Blood 2000; 96(4):1591-3.

71. Tavian M, Robin C, Coulombel L et al. The human embryo, but not its yolk sac, generates lympho-myeloid stem cells: Mapping multipotent hematopoietic cell fate in intraembryonic mesoderm. Immunity 2001; 15(3):487-95.

72. Palis J, Yoder MC. Yolk-sac hematopoiesis: The first blood cells of mouse and man. Exp Hematol 2001; 29(8):927-36.

CHAPTER 10

Hematopoietic Development in *Drosophila*:
A Parallel with Vertebrates

Marie Meister and Shubha Govind

Abstract

*D*rosophila hematopoiesis includes two distinct phases. The first phase occurs during the second half of embryonic development, whereas the second hematopoietic phase occurs during the larval stages. Blood cells (hemocytes) are responsible mainly for phagocytosis, encapsulation of large invaders, and associated humoral melanisation reactions. The primary *Drosophila* hemocyte lineage is essentially akin to the mammalian myeloid lineage. In this review we describe the current knowledge of the molecular mechanisms that govern hemocyte proliferation and lineage specification while drawing parallels with mammalian hematopoiesis. A genetic hierarchy of four transcription factors regulates lineage specification. Mammalian homologs of three of these four transcription factors also have critical functions in mammalian hematopoiesis. In addition, as in mammalian hematopoiesis, signaling pathways (Toll-NF-κB, JAK-STAT and steroid hormone) and chromatin remodeling complexes regulate specific aspects of proliferation and differentiation in *Drosophila* hematopoiesis.

Introduction

Hematopoiesis, the process of blood cell formation occurs in almost all animal phyla. Blood cells play important immune and nonimmune functions in these organisms. The process of hematopoiesis has been investigated in a number of nonmammalian species. In Arthropods, hematopoiesis has so far been described mainly at the morphological level only in a very few species. *Drosophila melanogaster*, one of the most popular animal models is an exception. After the pioneering studies of the Rizkis and E. Gateff,[85,94] research on *Drosophila* blood cells and hematopoiesis has recently attracted renewed interest.

In *Drosophila*, as in all other invertebrates investigated, host defense is based solely on innate immunity mechanisms, whereas vertebrates have combined innate and adaptive immune systems to defend themselves. Nevertheless, the immune system of invertebrates is extremely efficient and involves both humoral and cellular components. Over the last ten years, the molecular mechanisms underlying *Drosophila* humoral immunity have been the focus of intense research. Significantly, the mechanisms that control innate immune response in *Drosophila* and mammals are highly conserved as they both involve NF-κB- dependent activation cascades.[22,42,49]

Hematopoietic Stem Cell Development, edited by Isabelle Godin and Ana Cumano.
©2006 Eurekah.com and Kluwer Academic / Plenum Publishers.

Cellular immune responses are activated when *Drosophila* is wounded or challenged with microscopic or macroscopic infectious agents. Here, we briefly review the blood cell types, or hemocytes, that participate in cellular immunity, and we also provide an overview of our current knowledge of hematopoiesis in *Drosophila*, underlining the parallels with mammals.

Drosophila Hemocyte Types and Their Functions

In *Drosophila*, three mature hemocyte types are found in circulation, or hemolymph (the blood analogue of invertebrates): plasmatocytes, crystal cells and lamellocytes (Fig. 1).[54,85,94]

At the larval stage a majority of the cells (>90%), called *plasmatocytes*, exhibit strong phagocytic activity. At the onset of metamorphosis, they further differentiate into active macrophage-like cells and participate in tissue remodeling by ingesting doomed tissues. The plasmatocyte lineage is comparable, both in terms of ultrastructure and of function, to the mammalian monocyte/macrophage lineage. Present at all stages from the embryo to the adult, it represents the totality of the blood cell pool at the latter stage. The contribution of plasmatocytes to host defense is considerable. This is illustrated in experiments with larvae carrying reduced blood cells or flies in which blood cells have been inactivated. Such aberrant animals exhibit reduced survival to bacterial infection.[9,21]

In embryos and larvae, a smaller proportion of hemocytes are crystal cells, initially identified for their large cytoplasmic crystalline inclusions (Fig. 1A, C). The crystals were proposed to correspond to the enzymes and substrates responsible for humoral melanisation.[83] We showed that they contain at least the zymogen of the key enzyme in this process, namely prophenoloxidase (M. Meister, unpublished).

Hemolymph-borne melanisation is considered to be a defense mechanism common in invertebrates and presumably serves to fight parasites that enter the body.[68] Melanisation has no known equivalent in vertebrates. It is easily observed when a black capsule forms around an invader that is too large to be phagocytosed. Melanisation is also activated and has been found to be essential in a wound healing reaction where a black melanin smear rapidly appears at the wound site to seal it.[77] The activation of melanisation has been mostly studied in crustaceans[96] and in the silkworm *Bombyx mori*.[2] These studies have established that a serine protease cascade, which ultimately cleaves inactive prophenoloxidase into phenoloxidase, is activated upon recognition of foreign material. Phenoloxidase is an oxidoreductase that catalyses the oxidation of phenols to quinones. It thus catalyses several key steps in the conversion of tyrosine to the melanin. Despite the wide distribution of melanisation reactions in invertebrates, a clear role for melanisation in host defense has not been firmly established.

The third hemocyte type is only occasionally encountered in healthy animals and is called the *lamellocyte*. Lamellocytes differentiate after *Drosophila* larvae are infected by parasitic wasps (Fig. 1D). Since asp eggs are too large to be phagocytosed, lamellocytes wrap around the invader to form a capsule (Fig. 1E). Encapsulation is accompanied by melanisation, killing the parasite presumably by asphyxiation or through the local production of cytotoxic intermediates during melanin synthesis.[13] Lamellocytes are large flat cells, with few cytoplasmic organelles, and have never been observed at the embryonic and the adult stages.

Arthropods appear to lack cells equivalent to those in the mammalian lymphoid lineage. *Drosophila* plasmatocytes, as well as a number of Arthropod hemocytes investigated so far, are reminiscent both in terms of ultrastructure and of functions of vertebrate myeloid cells, namely the monocyte/macrophage lineage. It is not possible to relate crystal cells or lamellocytes to any of the mammalian blood cell types. These features make *Drosophila* a suitable model for the analysis of simple lineage specification and differentiation.

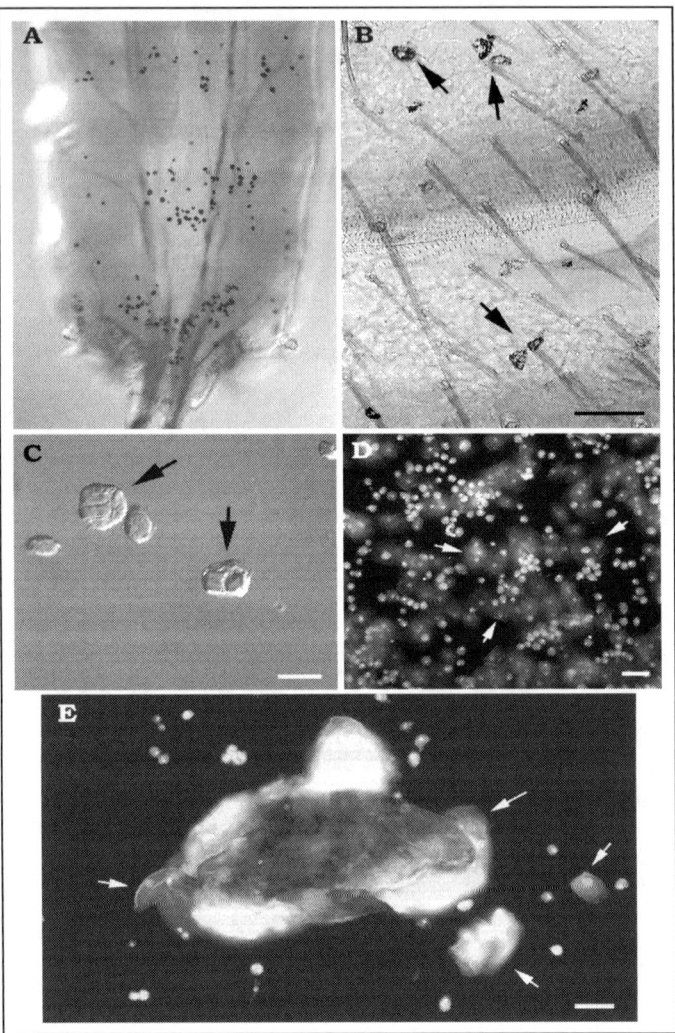

Figure 1. *Drosophila* hemocytes. A) Larval crystal cells visualized by heat-treatment at 65°C for 10 min: the cells melanize and are visible here through the transparent cuticle in the posterior part of the larva; B) Adult plasmatocytes visualized by injection of India ink into the fly: plasmatocytes take up the ink particles and are visible through the cuticle (arrows); C) Phase photomicrograph of larval crystal cells (black arrows); D) E) DAPI staining of hemocytes in circulation dissected from a larva 48 h after parasitization by *Leptopilina boulardi*. Note the presence of abundant lamellocytes (white arrows), and in E) of a parasite larva surrounded by lamellocytes. The small round cells are mainly plasmatocytes. Bars: 10 μm in C, 50 μm in B, D, E.

Hematopoietic Development

   Drosophila is a holometabolous insect, which implies that its post-embryonic development includes a period of complete metamorphosis that transforms a crawling larva into a flying adult. Four distinct developmental stages, embryo, larva, pupa and adult can be defined. Each stage presents specific hematopoietic features.

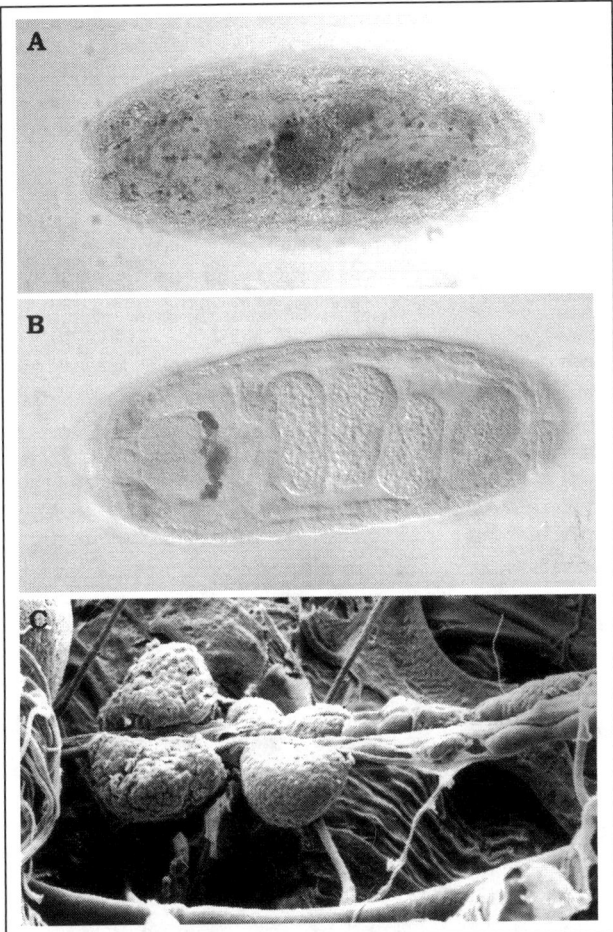

Figure 2. Hematopoiesis in embryonic and larval stages in *Drosophila*. A) Embryonic macrophages stained with anti-croquemort antibody;[25] B) embryonic crystal cells stained with anti-prophenoloxidase antibody;[66] C) Scanning electron microscopy of larval lymph glands: the lobes are paired and located along the dorsal vessel. The anterior-most lobes are the largest, and the hemocytes that make up the organ are clearly delineated. A-C: anterior to the left.

At the embryonic stage, hematopoiesis begins approximately at mid-development, when a population of cells expressing an early hemocyte marker (peroxidasin)[69] differentiates in the procephalic mesoderm.[99] These peroxidasin-positive cells disperse throughout the embryo. As they migrate, they become active macrophages which clear apoptotic cells within the developing embryo (Fig. 2A). Their phagocytic role is dependent on the wildtype function of the *croquemort* gene encoding a CD36 homologue.[25,26] A second hemocyte population appears at mid-embryogenesis which remains located around the anterior region of the gut and gives rise to embryonic crystal cells (Fig. 2B).[55] Finally, toward the end of embryogenesis, the future larval hematopoietic organ, the lymph gland, differentiates dorsally from clusters of lateral mesodermal cells.[88]

During the three *Drosophila* larval stages, the lymph glands are the major site of hemocyte production.[3,54,82,83,85,94] They are composed of several pairs of lobes located along the dorsal vessel, a pulsatile open structure that ensures a continuous stream of hemolymph throughout the body (Fig. 2C). Lymph glands contain a majority of small undifferentiated cells, prohemocytes, which are believed to be "multipotent" precursors. The anteriormost lobes contain numerous differentiated cells (plasmatocytes and crystal cells) and their progenitors while more posterior lobes mostly contain prohemocytes. Lymph glands also harbor a blood cell type, the secretory cells,[54,94] that was not found outside the hematopoietic organ. These secretory cells exhibit ultrastructural features characteristic of active protein synthesis, and their function is not yet understood. In the hemolymph, circulating cells include mature plasmatocytes and crystal cells,[85] but also a small proportion of prohemocytes. Further, a considerable number of hemocytes (about one third of the total larval blood cell pool) are nested in sessile islets that are attached to the epidermis.[54] The islets are segmentally distributed and located mostly at dorsal and posterior positions along the body wall. The blood cell composition of the sessile islets generally reflects that of circulating cells, except that it never includes lamellocytes even after infection.

At the onset of metamorphosis, all plasmatocytes change their adhesion properties and become so-called pupal macrophages.[54] This modification is under the control of ecdysone (see below), the steroid hormone that orchestrates moults and metamorphosis in insects.[100] Simultaneously in the lymph glands, an abundance of macrophages is observed in all lobes, including those which previously only contained prohemocytes. These macrophages ingest the remaining secretory cells as well as the basement material that delineates the lymph glands, before rejoining circulation. Thus, only a few hours after the onset of metamorphosis, the hematopoietic organ is no longer detectable. At later stages (in pupae and adults), there is no hematopoietic organ. The only blood cell type present in adults is the plasmatocyte, which maintains strong phagocytic activity (Fig. 1B). Mitotic activity has not been reported in the adult hemocyte pool. It is thus believed that adult hemocytes derive from larval/pupal blood cells.

Genetic Control of Hematopoiesis

A number of genes that control *Drosophila* hematopoiesis have been identified. Available data are derived both from observations of hematopoietic defects in *Drosophila* mutants or from the analysis of *Drosophila* counterparts of mammalian factors known to be essential in hematopoietic development. These genes encode transcription factors as well as components of signaling pathways and chromatin-remodelling proteins.

GATA Factor Serpent Specifies Hemocytes

Transcription factors of the GATA family are characterized by a highly conserved DNA binding domain containing two zinc fingers. In vertebrates, they are involved in differentiation and development of various organs. GATA-1, -2 and -3 are required for different aspects of both myeloid and lymphoid hematopoiesis[70] while GATA-4, -5 and -6 play a role in the formation of endoderm, lung, heart and genitourinary tract.[65]

Five GATA encoding genes have been described in *Drosophila*. Their expression patterns and functions have been investigated mainly in embryos. *pannier* or *dGATAa*, *grain* or *GATAc*, *dGATAd* and *e* have not been implicated in hematopoiesis. *serpent* (*srp*) or *dGATAb* is expressed mainly in five embryonic regions: midgut primordium, yolk, amnioserosa, fat body and hemocytes. In all these tissues, *srp* is required for proper development of the corresponding structures.[78,80,81] Significantly in *srp* mutant embryos, the early differentiation of prohemocytes is impaired as evidenced by the reduced size of the initial hemocyte primordium.[78] Moreover in *srp* embryos, no mature hemocytes are formed. It was proposed that *srp* might be required

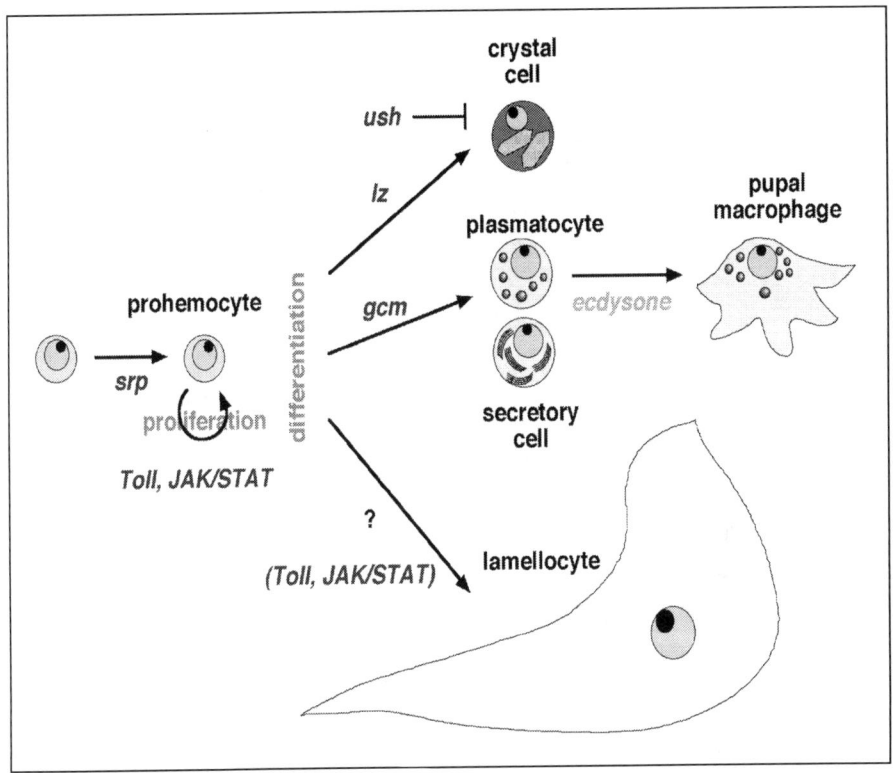

Figure 3. A model for hematopoiesis in *Drosophila* (modifed from ref. 54). The gene encoding the GATA factor Serpent (*srp*, also called *dGATAb*) is required for the specification of all blood cells. The transcription factors Glial Cells Missing (*gcm*) and Runt-family member Lozenge (*lz*) are required for the specification of plasmatocytes and crystal cells, respectively. The Friend-Of-GATA homologue U-shaped (*ush*) antagonizes the differentiation of crystal cells. The Toll-NF-κB and JAK/STAT pathways play a role in hemocyte proliferation and likely in lamellocyte specification. It is probable, although not demonstrated, that secretory cells and plasmatocytes are variants of one lineage as intermediate forms can be observed within the hematopoietic organ.

both for the specification of hemocyte primordium within the mesoderm at an early stage, and later for gene expression during their maturation.

It is thus clear that the GATA factor *srp* is indispensable in *Drosophila* for embryonic hematopoiesis. The *srp* mutations are late embryonic lethals. The role of *srp* has not been tested during the larval stages, but as SRP protein is expressed in all lymph gland cells in last instar larvae,[55] it is likely that its hematopoietic function is not restricted to the embryonic stage.

Lineage Specification by Transcription Factors

To date, three transcription factors have clearly been demonstrated to govern lineage specification in *Drosophila* hematopoietic development (Fig. 3). The glial cells missing (*glide/gcm*)[43,106] gene is expressed early and transiently during embryonic macrophage differentiation.[6] In loss-of-function *gcm* mutants, one third of the hemocytes are absent, and that those present exhibit defects in migration. Furthermore, ubiquitous ectopic *gcm* expression during embryonic development leads to a dramatic increase in peroxidasin-positive cells, which is not limited to the procephalic mesoderm territories. It was not known if these additional hemocytes were previously SRP-positive, or if *gcm* is a molecular target of SRP.

A role for the transcription factor Lozenge (LZ) in *Drosophila* hematopoiesis was investigated by Lebestky et al (2000). LZ shares 71% identity within its Runt DNA-binding domain, to that of mammalian AML1.[17] Interestingly, human AML1 is the most frequent target of chromosomal translocations in acute myeloid leukemias[76] and, like GATA-2, is essential for all definitive hematopoiesis in mouse.[107]

In *Drosophila* hematopoiesis, *lz* function is required for crystal cell production and differentiation during embryogenesis and later in larvae, as *lz* mutant embryos and larvae are devoid of this blood cell type.[55,84] As mentioned above, crystal cells in larvae originate in the lymph glands, and their precursors (CCPs) are LZ-positive. *srp* functions upstream of *lz* during embryogenesis, and likely also in larvae as *srp* expression is observed prior to that of *lz* in CCPs.

Banerjee and coworkers also showed that *gcm* is not expressed in embryonic CCPs. Consistent with this observation, *gcm* mutations do not affect *lz* expression in CCPs. However, misexpression of *gcm* in CCPs both in embryos and in larvae, specifies their change of fate to plasmatocytes and prevents differentiation of crystal cells. Based on these observations, the authors propose the following model: (i) a pool of SRP-positive cells serves as precursors for plasmatocytes and crystal cells; (ii) restricted expression of *gcm* is required for the developmental program of plasmatocytes; misexpression of *gcm* can override *lz*-mediated crystal cell specification and differentiation; (iii) restricted expression of *lz* specifies crystal cell differentiation. Strikingly, misexpression of *lz* in the entire hemocyte pool does not convert plasmatocytes into crystal cells.[55]

While GCM homologs have been described in mammals,[1,50] their expression in hematopoietic cells has not been observed and their role in hematopoietic development has not been reported so far.

Finally, a role for the third transcription factor, a Friend-Of-GATA (FOG) homolog, U-shaped (USH),[36] was recently discovered in hemocyte type specification of *Drosophila*.[23] FOG zinc-finger proteins regulate GATA factor activated gene transcription. During vertebrate hematopoiesis, FOG and GATA proteins cooperate to promote erythrocyte and megakaryocyte differentiation.[16,104] *Drosophila ush* is expressed in hemocyte precursors and plasmatocytes throughout embryonic and larval development[23,24] and *srp* expression is essential for embryonic *ush* expression. *ush* loss-of-function mutations result in overproduction of crystal cells and conversely, overexpression of *ush* reduces crystal cells number. *ush* thus functions as a negative regulator of crystal cell development.

Control of Lamellocyte Specification

Whereas the role of *gcm*, *lz* and *ush* is unambiguously established in the specification of plasmatocyte/macrophage and crystal cell lineages, factors that determine lamellocyte development are not as clearly defined. A number of mutations are known to result in aberrant lamellocyte differentiation. However, it is possible that in some of these mutants, lamellocyte differentiation is activated by nonhematopoietic factors. Mutations in three genes have long been known to stimulate lamellocyte production with formation of melanotic tumors in the hemocoel. These are *Toll* and JAK dominant gain-of-function, and *cactus* loss-of-function mutations. The role of the Toll and the JAK-STAT pathways in *Drosophila* hematopoiesis are outlined below and summarized in Figure 4. It is important to note that, like *srp*, *gcm* and *lz*, the components of the Toll and JAK/STAT pathways and other hematopoietic genes also have multiple roles during development.

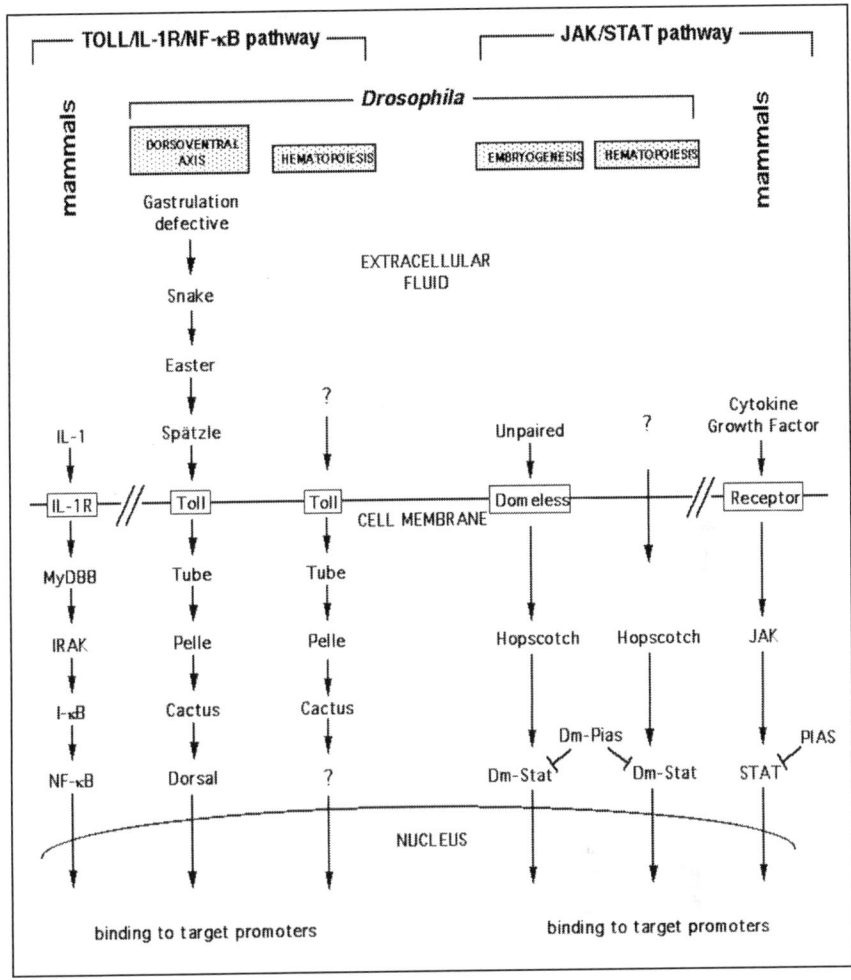

Figure 4. Comparative Toll/IL-1R/NF-κB and JAK/STAT signal transduction pathways in mammals and *Drosophila*. Mammalian IL-1R pathway is presented leftmost in a simplified manner as a representative of such pathways.[46] The proteins of the Toll pathway, as originally elucidated for the determination of dorsoventral polarity in *Drosophila* embryogenesis, are in the second column. The Toll-Cactus cassette has been implicated in governing *Drosophila* larval hemocyte proliferation. Members of the *Drosophila* JAK/STAT pathway, required for embryonic development, are in the fourth column. Hopscotch, Dm-Stat, and Dm-Pias have been implicated in proliferation and differentiation of hemocytes. The mammalian JAK/STAT pathway is rightmost.[45]

The Toll Pathway

The Toll pathway was initially described in the establishment of the dorso-ventral axis during early embryogenesis in *Drosophila*.[5,33] More recently, it was also shown to be responsible for activation of anti-fungal and anti-Gram positive response in the fat body of infected animals.[57,64] Toll receptors were subsequently identified in mammals where they serve as receptors for nonself, and control the immune response in several immunocompetent tissues/cells.[63] The ultimate activation step in the Toll pathway is the nuclear translocation of Rel transactivators that regulate multiple target genes. Rel proteins and their I-κB inhibitors have been implicated in hematopoietic processes in mammals through the analysis of knock-out mice phenotypes.[19,34] As an example, RelB and NF-κB were shown to play critical roles in hematopoietic system and lymphocyte development respectively; I-κBα knock-out results, among others, in atrophy of spleen and thymus, and abnormal B and T cell maturation.

In *Drosophila*, all the three Rel proteins control the induction of anti-microbial peptides in the fat body.[95] Of these, Dorsal (DL) and DIF (Dorsal-related Immunity Factor) are controlled by the Toll pathway, and Relish functions in a Toll-independent manner. A first indication that *Toll* and the related pathway is involved in blood cell development came from observations of the severe hematopoietic disorders in dominant gain-of-function alleles (*TollD*) in which this transmembrane receptor is constitutively active.[32,56] These mutants exhibit good viability and in larvae, the circulating hemocyte density is significantly higher (two-fold) than in wildtype. Moreover, >50% of these cells are lamellocytes. This phenotype frequently includes the formation of melanotic tumors, where the excess lamellocytes encapsulate self tissue within the hemocoel. The capsules generally blacken due to the simultaneous activation of crystal cells. As mentioned above, a comparable phenotype of hemocyte overproliferation (up to ten-fold), lamellocyte differentiation and melanotic tumor formation is found in strong loss-of-function alleles of *cactus* (*cact*).[74,87] *cact* encodes the unique *Drosophila* I-κB homolog.

These hematopoietic disorders in both *TollD* and *cact* mutants prompted an analysis of the role of the intracellular components of the related cascade in *Drosophila* hematopoietic development.[74] It was shown that in loss-of-function mutants for *Toll, tube* and *pelle* (*tube* encodes a death-domain adaptor protein and *pelle* a protein kinase), the hemocyte count in circulation is significantly lowered. Toll, Tube and Pelle, as well as Cact, Dif and Dl are all expressed in the lymph glands and represent potential players in the control of blood cell proliferation in *Drosophila* larvae.

The Toll pathway is thus involved in the control of hemocyte density in *Drosophila* larvae. The abnormal production of lamellocytes in *TollD* and *cact* loss-of-function mutants could have reflected an indirect effect possibly due to defects in fat body cells that are detected as abnormal self by overabundant hemocytes. However, experimental evidence speaks against this possibility as *cact* rescue experiments with a UAS-CACT transgene using hematopoietic GAL4 drivers that do not have fat body expression.[74]

The JAK/STAT Pathway

In mammals, a number of different JAK kinases and associated STAT and PIAS partners are responsive to cytokines and growth factors to transduce signals for the proliferation and differentiation of various cell types, particularly in hematopoietic lineages.[45] In contrast, a single JAK-STAT pair is encoded in the *Drosophila* genome. The role of the JAK gene (named *hopscotch* or *hop*) was first documented in embryonic development where it regulates segmentation by controling pair-rule gene transcription.[8,71] It turned out that a severe hematopoietic disorder in *Drosophila37* is due to a dominant mutation in the *hop* gene product, where an amino-acid substitution generates a constitutively active form of the *Drosophila* JAK kinase (*hopTum-l*).[38,60] This mutation results in dramatic overproliferation of larval hemocytes, overgrowth of the hematopoietic organs, abnormal differentiation of lamellocytes accompanied by the formation of abundant melanotic tumors. Transplantation of mutant lymph glands into the abdomen of wildtype adults resulted in rapid death of the host presumably due to invasiveness of the cells from the transplanted lymph glands.[37] *hop* is ubiquitously required for cell proliferation in *Drosophila* as loss-of-function alleles result in underproliferation in all diploid tissues in larvae.[38] However, its role in hemocyte proliferation is particular, for in *hopTum-l* larvae, only blood cells exhibit the overproliferation phenotype.

A screen for mutations that suppress dominant hematopoietic defects of *hopTum-l* led to the identification of *Drosophila* STAT.[44,61,109] Significantly, a reduction in *Dm-stat* function does not modify hemocyte overproliferation, but suppresses lamellocyte differentiation and formation of melanotic tumors. Therefore Luo et al (1997) have proposed that *hop* controls both hemocyte proliferation and lamellocyte differentiation in *Drosophila* larvae; *Dm-stat* is required only for lamellocyte differentiation. These data are however still controversial as different results were obtained by Remillieux and coworkers[79] who showed that *Dm-stat* is required for hemocyte proliferation in *hopTum-l* mutants.

The product of a *Drosophila pias* gene (PIAS for Protein Inhibitor of Activated Stats,[59] also appears to play a role in hematopoiesis. Loss-of-function mutations in *Dm-pias* increase the incidence of melanotic tumor formation in a *hopTum-l* background,[7] whereas its overexpression has the opposite effect. This indicates that in *Drosophila*, a PIAS homologue that acts downstream of *hop* is likely to participate in hemocyte regulation.

The receptor(s)/ligand(s) that act usptream of the JAK-STAT pathway have long remained elusive in *Drosophila*. In particular, no sequence with clear homology with mammalian cytokines (e.g., interferon or interleukins) is predicted from the genome sequence. However, a putative ligand was proposed based on the embryonic phenotype of the corresponding mutants. The *unpaired* gene (*upd*) encodes a secreted protein, associated with the extracellular matrix, that activates the JAK pathway in embryogenesis[39] and in eye development.[110] While UPD and vertebrate cytokines do not share sequence similarity, it is intriguing that the 3D structure of UPD contains two α-helical regions reminiscent of cytokine structure.[39]

The identification of a putative receptor acting upstream of the JAK-STAT pathway is even more recent. domeless/master of marelle (dome/mom) encodes a transmembrane protein required for all JAK-STAT functions in the embryo.[12,14] Despite extensive sequence divergence, DOME/MOM has a significant similarity to the vertebrate gp130 family of cytokine receptors. The hypothesis that UPD is the ligand of DOME/MOM is supported both by genetic interaction experiments, and by direct interaction of proteins in transfected cells. The role of both *upd* and *dome/mom* in hematopoiesis is not yet known.

Hematopoiesis and Chromatin Remodelling

A large number of chromatin remodeling complexes have been identified in all eukaryotes from yeast to human.[51] Such complexes exert either repressive or activating regulatory activities on gene transcription, depending on the target genes, on the tissue, the developmental stage. In *Drosophila*, two groups of genes encoding subunits of such complexes have been extensively analysed. They control *Hox* gene expression in development.[48] Genetic interaction studies have defined the Polycomb group (PcG) proteins as repressors and trithorax group (trxG) proteins as activators of *Hox* gene activity. A number of *PcG* and *trxG* genes were identified in *Drosophila* and, subsequently, in vertebrates where they also play a major role in ensuring correct development.[11] It was shown that not only do these proteins maintain proper cell identity, but they also contribute to the regulation of cell proliferation.

Hox genes carry out important functions in the organisation and regulation of hematopoiesis in mammals.[105] It could thus be expected that their regulators also contribute to the control of hematopoietic processes. Indeed, a role for *PcG* genes in hematopoiesis was demonstrated in mice carrying mutations in several *PcG* genes. These mice exhibit various hematological disorders ranging from impaired blood cell proliferation to loss of mature B and T cells.[75,98] Furthermore, the human *trithorax* homologue MLL/ALL1[35] is a frequent target of acute leukemia-associated translocations.[15,67] It is likely that mammalian *PcG*, and possibly also *trxG* genes, control hematopoietic proliferation and lineage specification.

In *Drosophila*, a member of the PcG plays a crucial role in control of hemocyte proliferation. Loss-of-function mutations of *mxc* (*multi sex combs*;[91] result in severe lymph gland overgrowth, hyperproliferation of hemocytes and formation of melanotic tumors.[28,79,90] Although *mxc* mutations cause pleiotropic defects, the blood cell phenotype is cell-autonomous, as mutant lymph glands exhibit invasive properties when transplanted into a wildtype host.[30,79] Interestingly, the hyperproliferation of blood cells in *mxc* mutants is Toll pathway-dependent (as it is suppressed in loss-of-function mutations of *Toll* and *tube*), but is STAT-independent.[79]

Combination of *mxc* alleles with other *PcG* mutations does not result in additional effect on *mxc* hematopoietic phenotype, as might be expected from other developmental processes where *PcG* genes often act in synergy. Significantly however, decrease in wildtype function of the *trxG* gene *brahma* (which encodes a SWI2/SNF2 protein), had an antagonistic effect on *mxc* hemocyte hyperproliferation. These data suggest that *PcG* and *trxG* genes play a role in *Drosophila* hematopoiesis and provide additional facets of molecular mechanisms shared between *Drosophila* and mammals.

The *Drosophila domino* (*dom*) gene encodes two members of the SWI2/SNF2 family of DNA-dependent ATPases.[89] It can however not be classified among PcG or trxG genes as *dom* mutants do not exhibit homeotic transformations. DOM proteins are proposed to act as transcriptional repressors within large protein complexes that are believed to function by interfering with chromatin structure. The molecular nature of these complexes is not yet known. Loss-of-function mutations of *dom* result in ubiquitous proliferation defects and mutant *dom* larvae are devoid of circulating blood cells. The most striking phenotype is a melanisation of the hematopoietic organs due to cell death within the lymph glands.[9,10] Blood cell precursors fail to proliferate, then die during larval stages and blacken. Larvae carrying weak *dom* alleles show reduced proliferation within the lymph glands, but such hemocytes are abnormal and do not reach circulation. It is proposed that *dom* is necessary for blood cell proliferation in *Drosophila*. It is possibly also required for subsequent specification of blood cell identity. A human homologue for *dom* has been identified which is called SRCAP (SNF2-related CPB activator protein)[47] but its role in hematopoiesis has not been described so far.

A number of additional hemocyte proliferation mutants have been described in *Drosophila* larvae (Table 1).[18] Genes that are mutated encode ribosomal proteins (*air8*),[103,108] a protein necessary for chromosome condensation (proliferation disrupter),[103] a protein which can both participate in ribonucleoprotein formation and associate with chromatin (*modulo*),[72,73] an importin homologue (*oho31* or *pendulin*),[53,102] and a protein with homology to cytokeratins (*l(3) malignant blood neoplasm*).[52] The exact roles of these genes in hematopoiesis are not known.

A Drosophila PDGF/VEGF Receptor

Vascular endothelial growth factors or VEGFs in mammals play a central role in vasculogenesis and angiogenesis,[62] as they are mitogens highly specific for endothelial cells. The mammalian VEGF family currently includes 6 members and several of them exist as different isoforms that appear to have unique biological functions. They bind to three structurally related receptor tyrosine kinases or VEGF receptors. VEGF signaling has been proposed to function in hematopoiesis based on VEGFR-2 expression in hematopoietic stem cells.[111] In addition, VEGF-R1 plays a role in the migration of cells from the monocyte-macrophage lineage in mammals (see ref. 92 and references therein).

Insects have no endothelial cells or blood vessels: hemolymph fills the whole hemocoel and surrounds all organs. Despite the absence of an endothelial blood vessel system in *Drosophila*, the existence of three genes encoding VEGF homologues and one related receptor tyrosine kinase was recently reported.[20,41] The ligands were named PVF1-3 (for PDGF/VEGF Factor) and the receptor is called PVR (PDGF/VEGF receptor). PVF1 and PVR play a role in guidance of cell migration during *Drosophila* oogenesis.[20] Strikingly, the major splice form of PVR is expressed in cells of the embryonic macrophage lineage,[41] starting in the head mesoderm at mid-embryogenesis, then persisting as the cells migrate throughout the embryo. In embryos

Table 1. *Mutations that alter hematopoietic proliferation in* Drosophila *larvae*

Drosophila Gene	Protein	References	Vertebrate Homologues
Reduction in hematopoietic tissue			
domino	member of SWI2/SNF2 family	10, 89	SRCAP
l(3)hem	not characterized	29	
proliferation disrupter	chromatin binding	103	
Toll (g.o.f.)	transmembrane receptor	74, 40	TLRs, IL-1R
tube	death domain containing	74, 58	
pelle	serine/threonine kinase	74, 93	IRAK
Increase in hematopoietic tissue			
cactus	ankyrin-repeat containing	74, 31	I-κB
ecdysoneless	not characterized	27, 97	
hopscotch (g.o.f.)	receptor-associated tyrosine kinase	38, 60	Janus Kinase
l(3)mbn-1	cytokeratin domain containing	52	
multi sex combs	not characterized	91, 79	
oho23B	ribosomal protein	101	rpS21
pendulin	nuclear-localization-signal-binding protein	102, 53	importin
modulo	nucleolar; RNA binding domains	72, 73	
air8	ribosomal protein	108	rpS6
Toll (gain-of-function)	(see above)	74, 40, 56	

A list of *Drosophila* hematopoietic mutants organized by phenotype: under- or over-proliferation of hematopoietic tissue in larvae. Unless otherwise specified, mutations are loss-of-function mutations. g.o.f.: gain-of-function.

lacking a functional PVR gene, hemocytes are still present but their migration is hampered and they remain mainly located in the anterior region. The distribution of the three PVF proteins is not known at this stage. In the *Drosophila* S2 blood cell line, the addition of anti-PVR antibody to the culture medium abolishes cell division in conditions that normally stimulate mitosis (A. Munier & M. Lagueux, manuscript in preparation). This same antibody stains larval hemocyte sub-populations. These data indicate that a PVF/PVR signaling is used in *Drosophila* both for hemocyte migration in embryos, and probably for their proliferation at the larval stage.

Hormonal Control of Hematopoiesis

Larval and pupal development in *Drosophila* are under the tight control of hormones. As mentioned above, the steroid hormone ecdysone controls larval transitions (molts) and orchestrates metamorphosis, the conversion of the larval body into adult structures. In its molecular mechanism, the signal transduction pathway activated by ecdysone is very similar to that found in mammals: a complex of cytoplasmic DNA-binding hormone receptors is recruited to the nucleus.

By controlling the transcription of specific genes, these transcription factors unleash a cascade of molecular and cellular changes, that ultimately results in development of the adult fly.[100]

Like all other tissues of the larva, the lymph gland is under the control of ecdysone. Studies with the *ecdysoneless* mutants have revealed that the development of the hemocyte progenitors in the lymph gland is under hormonal control. *ecdysoneless1*, or *ecd1*, is a temperature sensitive mutant in which levels of ecdysone are very low at nonpermissive temperature.[27] As a result, lymph glands from third instar mutant larvae, while normal in morphology, are unable to respond to the parasitization. A limited burst of mitosis and the differentiation of both lamellocytes and crystal cells, changes that normally follow shortly after infection and are required for encapsulation, are blocked in *ecd* mutants.[97] Thus, an ecdysone-activated pathway appears to potentiate precursors of effector cell types to respond to parasitization by proliferating and differentiating. By affecting a specific pool of hematopoietic precursors, this pathway thus confers immune capacity to third instar larvae.

The requirement of ecdysone for hematopoiesis in normal and infected larvae is much like the role of glucocorticoid in mammalian erythropoiesis. Mice lacking the glucocorticoid receptor show no obvious defects in normal erythropoiesis. However, like infected *ecd1* mutant larvae, GR-deficient mice are unable to respond to stress situations such as erythrolysis or hypoxia and are unable to expand the population of immature erythroid cells during stress erythropoiesis.[4]

A second and distinct requirement for ecdysone in hematopoiesis is evident at the larval-to-pupal transition. As mentioned above, the appearance of macrophages in the lymph gland coincides with the onset of pupariation. In a few hours subsequent to this transition, the lymph gland is no longer detectable.[54] At nonpermissive temperature, in developmentally-delayed *ecd1* larvae, lymph gland integrity is maintained ; unable to make the transition, they do not disperse but continue to grow and become hypertrophic.[97]

Drosophila as a Model System for Understanding Mammalian Hematopoiesis

For a long time *Drosophila* has failed to be a leading model organism in the analysis of hematopoietic mechanisms, despite its long-standing importance in the study of other developmental processes. However, a number of recent studies described in this review, have demonstrated that *Drosophila* has a role to play in the understanding of blood cell development. In particular, the function of transactivators such as GATA, FOG and AML-1 homologues, of pathways such as the Toll-NF-κB and the JAK-STAT signaling cassettes, have been successfully investigated in *Drosophila* hematopoiesis, and clear parallels with mammalian mechanisms can be drawn. In addition, evidence for the involvement of other *Drosophila* genes such as *gcm* or *mxc* for instance, could provide new directions to research in mammalian hematopoiesis.

The lack of data previously was due mostly to the absence of cell markers in *Drosophila*. While an increasing number of markers for mature hemocytes are now available, markers for multipotent stem cells and for precursors of the various lineages are still lacking. Future data on *Drosophila* blood cell development and functions should accumulate rapidly, and thanks to the powerful genetics of this model system, soon contribute to our general understanding of hematopoietic processes and blood cell functions in invertebrates and vertebrates.

Acknowledgements

Because of lack of space, all primary references could not be included in this article. We would like to thank Zakaria Kambris, Julien Royet and Daniel Zachary for providing figures. Research in our laboratories was funded by grants from the Centre National de la Recherche Scientifique, the Association pour la Recherche sur le Cancer and the Ligue Nationale et Régionale contre le Cancer (MM), and the American Heart Association, Heritage Affiliate, American Cancer Society, NIH-RCMI RR03060-16 and PSC-CUNY (SG).

References

1. Akiyama Y, Hosoya T, Poole AM et al. The gcm-motif: A novel DNA-binding motif conserved in Drosophila and mammals. Proc Natl Acad Sci USA 1996; 93:14912-6.

2. Ashida M, Brey PT. Recent advances in research on the insect propheno-loxidase cascade. In: Brey PT, Hultmark D, eds. Molecular Mechanisms of Immune Response in insects. London: Chapman & Hall, 1997:135-172.

3. Bairati A. L'ultrastruttura dell'organo dell'emolinfa nella larva di Drosophila melanogaster. Zeitschrift für Zellforschung 1964; 61:769-802.

4. Bauer A, Tronche F, Wessely O et al. The glucocorticoid receptor is required for stress erythropoiesis. Genes Dev 1999; 13:2996-3002.

5. Belvin MP, Anderson KV. A conserved signaling pathway: The Drosophila toll-dorsal pathway. Annu Rev Cell Dev Biol 1996; 12:393-416.

6. Bernardoni R, Vivancos B, Giangrande A. Glide/gcm is expressed and required in the scavenger cell lineage. Dev Biol 1997; 191:118-30.

7. Betz A, Lampen N, Martinek S. A Drosophila PIAS homologue negatively regulates stat92E. Proc Natl Acad Sci USA 2001; 98: 9563-8.

8. Binari R, Perrimon N. Stripe-specific regulation of pair-rule genes by hopscotch, a putative Jak family tyrosine kinase in Drosophila. Genes Dev 1994; 8:300-12.

9. Braun A, Hoffmann JA, Meister M. Analysis of the Drosophila host defense in domino mutant larvae, which are devoid of hemocytes. Proc Natl Acad Sci USA 1998; 95:14337-42.

10. Braun A, Lemaitre B, Lanot R et al. Drosophila immunity: Analysis of larval hemocytes by P-element-mediated enhancer trap. Genetics 1997; 147:623-34.

11. Brock HW, van Lohuizen M. The Polycomb group—no longer an exclusive club? Curr Opin Genet Dev 2001; 11:175-81.

12. Brown S, Hu N, Hombria JC. Identification of the first invertebrate interleukin JAK/STAT receptor, the Drosophila gene domeless. Curr Biol 2001; 11:1700-5.

13. Carton Y, Nappi AJ. Drosophila cellular immunity against parasitoids. Parasitology Today 1997; 13:218-227.

14. Chen HW, Chen X, Oh SW et al. Mom identifies a receptor for the Drosophila JAK/STAT signal transduction pathway and encodes a protein distantly related to the mammalian cytokine receptor family. Genes Dev 2002; 16:388-98.

15. Cimino G, Rapanotti MC, Sprovieri T et al. ALL1 gene alterations in acute leukemia: Biological and clinical aspects. Haematologica 1998; 83:350-7.

16. Crispino JD, Lodish MB, MacKay JP et al. Use of altered specificity mutants to probe a specific protein-protein interaction in differentiation: The GATA-1:FOG complex. Mol Cell 1999; 3:219-28.

17. Daga A, Karlovich CA, Dumstrei K et al. Patterning of cells in the Drosophila eye by Lozenge, which shares homologous domains with AML1. Genes Dev 1996; 10:1194-205.

18. Dearolf CR. Fruit fly "leukemia." Biochim Biophys Acta 1998; 1377:M13-23.

19. Denk A, Wirth T, Baumann B. NF-kappaB transcription factors: Critical regulators of hematopoiesis and neuronal survival. Cytokine Growth Factor Rev 2000; 11:303-20.

20. Duchek P, Somogyi K, Jekely G et al. Guidance of cell migration by the Drosophila PDGF/VEGF receptor. Cell 2001; 107:17-26.

21. Elrod-Erickson M, Mishra S, Schneider D. Interactions between the cellular and humoral immune responses in Drosophila. Curr Biol 2000; 10:781-4.

22. Engstrom Y. Induction and regulation of antimicrobial peptides in Drosophila. Dev Comp Immunol 1999; 23:345-58.
23. Fossett N, Tevosian SG, Gajewski K et al. The Friend of GATA proteins U-shaped, FOG-1, and FOG-2 function as negative regulators of blood, heart, and eye development in Drosophila. Proc Natl Acad Sci USA 2001; 98:7342-7.
24. Fossett N, Zhang Q, Gajewski K et al. The multitype zinc-finger protein U-shaped functions in heart cell specification in the Drosophila embryo. Proc Natl Acad Sci USA 2000; 97:7348-53.
25. Franc NC, Dimarcq JL, Lagueux M et al. Croquemort, a novel Drosophila hemocyte/macrophage receptor that recognizes apoptotic cells. Immunity 1996; 4:431-43.
26. Franc NC, Heitzler P, Ezekowitz RA et al. Requirement for croquemort in phagocytosis of apoptotic cells in Drosophila. Science 1999; 284:1991-4.
27. Garen A, Kauvar L, Lepesant JA. Roles of ecdysone in Drosophila development. Proc Natl Acad Sci USA 1977; 74:5099-5103.
28. Gateff E. Malignant neoplasms of the hematopoietic system in three mutants of Drosophila melanogaster. Ann Parasitol Hum Comp 1977; 52:81-3.
29. Gateff E. Tumor suppressor and overgrowth suppressor genes of Drosophila melanogaster: Developmental aspects. Int J Dev Biol 1994; 38:565-90.
30. Gateff E, Mechler BM. Tumor-suppressor genes of Drosophila melanogaster. Crit Rev Oncog 1989; 1:221-45.
31. Geisler R, Bergmann A, Hiromi Y et al. Cactus, a gene involved in dorsoventral pattern formation of Drosophila, is related to the I kappa B gene family of vertebrates. Cell 1992; 71:613-21.
32. Gerttula S, Jin YS, Anderson KV. Zygotic expression and activity of the Drosophila Toll gene, a gene required maternally for embryonic dorsal-ventral pattern formation. Genetics 1988; 119:123-33.
33. Govind S. Control of development and immunity by rel transcription factors in Drosophila. Oncogene 1999; 18:6875-87.
34. Grossmann M, Nakamura Y, Grumont R et al. New insights into the roles of ReL/NF-kappa B transcription factors in immune function, hemopoiesis and human disease. Int J Biochem Cell Biol 1999; 31:1209-19.
35. Gu Y, Nakamura T, Alder H et al. The t(4;11) chromosome translocation of human acute leukemias fuses the ALL-1 gene, related to Drosophila trithorax, to the AF-4 gene. Cell 1992; 71:701-8.
36. Haenlin M, Cubadda Y, Blondeau F et al. Transcriptional activity of pannier is regulated negatively by heterodimerization of the GATA DNA-binding domain with a cofactor encoded by the u-shaped gene of Drosophila. Genes Dev 1997; 11:3096-108.
37. Hanratty WP, Ryerse JS. A genetic melanotic neoplasm of Drosophila melanogaster. Dev Biol 1981; 83:238-49.
38. Harrison DA, Binari R, Stines Nahreini T et al. Activation of a Drosophila Janus kinase (JAK) causes hematopoietic neoplasia and developmental defects. EMBO J 1995; 14:2857-2865.
39. Harrison DA, McCoon PE, Binari R et al. Drosophila unpaired encodes a secreted protein that activates the JAK signaling pathway. Genes Dev 1998; 12:3252-63.
40. Hashimoto C, Hudson KL, Anderson KV. The Toll gene of Drosophila, required for dorsal-ventral embryonic polarity, appears to encode a transmembrane protein. Cell 1988; 52:269-79.
41. Heino TI, Karpanen T, Wahlstrom G et al. The Drosophila VEGF receptor homolog is expressed in hemocytes. Mech Dev 2001; 109:69-77.
42. Hoffmann JA, Kafatos FC, Janeway CA et al. Phylogenetic perspectives in innate immunity. Science 1999; 284:1313-8.
43. Hosoya T, Takizawa K, Nitta K et al. Glial cells missing: A binary switch between neuronal and glial determination in Drosophila. Cell 1995; 82:1025-36.
44. Hou XS, Melnick MB, Perrimon N. Marelle acts downstream of the Drosophila HOP/JAK kinase and encodes a protein similar to the mammalian STATs. Cell 1996; 84:411-9.
45. Ihle JN, Nosaka T, Thierfelder W et al. Jaks and Stats in cytokine signaling. Stem Cells 1997; 15 Suppl 1:105-11 discussion 112.
46. Imler JL, Hoffmann JA Signaling mechanisms in the antimicrobial host defense of Drosophila. Curr Opin Microbiol 2000; 3:16-22.

47. Johnston H, Kneer J, Chackalaparampil I et al. Identification of a novel SNF2/SWI2 protein family member, SRCAP, which interacts with CREB-binding protein. J Biol Chem 1999; 274:16370-6.

48. Kennison JA. The Polycomb and trithorax group proteins of Drosophila: Trans-regulators of homeotic gene function. Annu Rev Genet 1995; 29:289-303.

49. Khush RS, Leulier F, Lemaitre B. Drosophila immunity: Two paths to NF-kappaB. Trends Immunol 2001; 22:260-4.

50. Kim J, Jones BW, Zock C et al. Isolation and characterization of mammalian homologs of the Drosophila gene glial cells missing. Proc Natl Acad Sci USA 1998; 95:12364-9.

51. Kingston RE, Narlikar GJ. ATP-dependent remodeling and acetylation as regulators of chromatin fluidity. Genes Dev 1999; 13:2339-52.

52. Konrad L, Becker G, Schmidt A et al. Cloning, structure, cellular localization, and possible function of the tumor suppressor gene lethal(3)malignant blood neoplasm-1 of Drosophila melanogaster. Dev Biol 1994; 163:98-111.

53. Kussel P, Frasch M. Pendulin a drosophila protein with cell cycle-dependent nuclear localization, is required for normal cell proliferation. J Cell Biol 1995; 129:1491-507.

54. Lanot R, Zachary D, Holder F et al. Postembryonic hematopoiesis in Drosophila. Dev Biol 2001; 230:243-57.

55. Lebestky T, Chang T, Hartenstein V et al. Specification of Drosophila hematopoietic lineage by conserved transcription factors. Science 2000; 288:146-9.

56. Lemaitre B, Meister M, Govind S et al. Functional analysis and regulation of nuclear import of dorsal during the immune response in Drosophila. EMBO J 1995; 14:536-45.

57. Lemaitre B, Reichhart JM, Hoffmann JA. Drosophila host defense: Differential induction of antimicrobial peptide genes after infection by various classes of microorganisms. Proc Natl Acad Sci USA 1997; 94:14614-9.

58. Letsou A, Alexander S, Orth K et al. Genetic and molecular characterization of tube, a Drosophila gene maternally required for embryonic dorsoventral polarity. Proc Natl Acad Sci USA 1991; 88:810-4.

59. Liao J, Fu Y, Shuai K. Distinct roles of the NH2- and COOH-terminal domains of the protein inhibitor of activated signal transducer and activator of transcription (STAT1) (PIAS1) in cytokine-induced PIAS1-Stat1 interaction. Proc Natl Acad Sci USA 2000; 97:5267-72.

60. Luo H, Hanratty WP, Dearolf CR. An amino acid substitution in the Drosophila hopTum-l Jak kinase causes leukemia-like hematopoietic defects. EMBO J 1995; 14:1412-20.

61. Luo H, Rose P, Barber D et al. Mutation in the Jak kinase JH2 domain hyperactivates Drosophila and mammalian Jak-Stat pathways. Mol Cell Biol 1997; 17:562-71.

62. Matsumoto T, Claesson-Welsh L et al. VEGF receptor signal transduction. Sci STKE 2001; RE21.

63. Medzhitov R, Janeway Jr C. The Toll receptor family and microbial recognition. Trends Microbiol 2000; 8:452-6.

64. Michel T, Reichhart JM, Hoffmann JA. Drosophila Toll is activated by Gram-positive bacteria through a circulating peptidoglycan recognition protein. Nature 2001; 414:756-759.

65. Molkentin JD. The zinc finger-containing transcription factors GATA-4, -5, and -6. Ubiquitously expressed regulators of tissue-specific gene expression. J Biol Chem 2000; 275:38949-52.

66. Muller HM, Dimopoulos G, Blass C et al. A hemocyte-like cell line established from the malaria vector Anopheles gambiae expresses six prophenoloxidase genes. J Biol Chem 1999; 274:11727-35.

67. Muller I, Aepinus C, Beck R et al. Noncutaneous varicella-zoster virus (VZV) infection with fatal liver failure in a child with acute lymphoblastic leukemia (ALL). Med Pediatr Oncol 2001; 37:145-7.

68. Nappi AJ, Vass E. Melanogenesis and the generation of cytotoxic molecules during insect cellular immune reactions. Pigment Cell Res 1993; 6:117-26.

69. Nelson RE, Fessler LI, Takagi Y et al. Peroxidasin: A novel enzyme-matrix protein of Drosophila development. Embo J 1994; 13:3438-47.

70. Orkin SH. Embryonic stem cells and transgenic mice in the study of hematopoiesis. Int J Dev Biol 1998; 42:927-34.

71. Perrimon N, Mahowald AP. l(1)hopscotch, A larval-pupal zygotic lethal with a specific maternal effect on segmentation in Drosophila. Dev Biol 1986; 118:28-41.

72. Perrin L, Demakova O, Fanti L et al. Dynamics of the sub-nuclear distribution of Modulo and the regulation of position-effect variegation by nucleolus in Drosophila. J Cell Sci 1998; 111(Pt 18):2753-61.

73. Perrin L, Romby P, Laurenti P et al. The Drosophila modifier of variegation modulo gene product binds specific RNA sequences at the nucleolus and interacts with DNA and chromatin in a phosphorylation-dependent manner. J Biol Chem 1999; 274:6315-23.

74. Qiu P, Pan PC, Govind S. A role for the Drosophila Toll/Cactus pathway in larval hematopoiesis. Development 1998; 125:1909-20.

75. Raaphorst FM, Otte AP, Meijer CJ. Polycomb-group genes as regulators of mammalian lymphopoiesis. Trends Immunol 2001; 22:682-90.

76. Rabbitts TH. Chromosomal translocations in human cancer. Nature 1994; 372:143-9.

77. Ramet M, Lanot R, Zachary D et al. JNK signaling pathway is required for efficient wound healing in Drosophila. Dev Biol 2002; 241:145-56.

78. Rehorn KP, Thelen H, Michelson AM et al. A molecular aspect of hematopoiesis and endoderm development common to vertebrates and Drosophila. Development 1996; 122:4023-31.

79. Remillieux-Leschelle N, Santamaria P, Randsholt N. Regulation of larval hematopoiesis in Drosophila melanogaster: A role for the multi sex combs gene. Genetics 2002; 162(3):1259-74.

80. Reuter R. The gene serpent has homeotic properties and specifies endoderm versus ectoderm within the Drosophila gut. Development 1994; 120:1123-35.

81. Riechmann V, Rehorn KP, Reuter R et al. The genetic control of the distinction between fat body and gonadal mesoderm in Drosophila. Development 1998; 125:713-23.

82. Rizki TM. The circulatory system and associated cells and tissues. In: Ashburner M,Wright TRF eds. Genetics and Biology of Drosophila. New York: Academic Press, 1978:2b:397-452.

83. Rizki TM, Rizki RM. Properties of the larval hemocytes of Drosophila melanogaster. Experientia 1980; 36:1223-1226.

84. Rizki TM, Rizki RM. Alleles of lz as suppressors of the Bc phenotype in Drosophila melanogaster. Genetics 1981; 97:s90.

85. Rizki TM, Rizki RM. The cellular defense system of Drosophila melanogaster. In: King RC, Akai H eds. Insect ultrastructure. New York: Plenum Publishing Corporation, 1984:2:579-604.

86. Rizki TM, Rizki RM, Grell EH. A mutant affecting the crystal cells in Drosophila melanogaster. Wilhelm's Roux's Archives 1980; 188:91-99.

87. Roth S, Hiromi Y, Godt D et al. cactus, a maternal gene required for proper formation of the dorsoventral morphogen gradient in Drosophila embryos. Development 1991; 112:371-88.

88. Rugendorff A, Younossi-Hartenstein A, Hartenstein V. Embryonic origin and differentiation of the Drosophila heart. Roux's Archives of Developmental Biology 1994; 203:266-280.

89. Ruhf ML, Braun A, Papoulas O et al. The domino gene of Drosophila encodes novel members of the SWI2/SNF2 family of DNA-dependent ATPases, which contribute to the silencing of homeotic genes. Development 2001; 128:1429-41.

90. Saget O, Forquignon F, Santamaria P et al. Needs and targets for the multi sex combs gene product in Drosophila melanogaster. Genetics 1998; 149:1823-38.

91. Santamaria P, Randsholt NB. Characterization of a region of the X chromosome of Drosophila including multi sex combs (mxc), a Polycomb group gene which also functions as a tumour suppressor. Mol Gen Genet 1995; 246:282-90.

92. Sawano A, Iwai S, Sakurai Y. Flt-1, vascular endothelial growth factor receptor 1, is a novel cell surface marker for the lineage of monocyte-macrophages in humans. Blood 2001; 97:785-91.

93. Shelton CA, Wasserman SA. Pelle encodes a protein kinase required to establish dorsoventral polarity in the Drosophila embryo. Cell 1993; 72:515-25.

94. Shrestha R, Gateff E. Ultrastructure and cytochemistry of the cell types in the larval hematopoietic organs and hemolymph of Drosophila melanogaster. Dev Growth Differ 1982; 24:65-82.

95. Silverman N, Maniatis T. NF-kappaB signaling pathways in mammalian and insect innate immunity. Genes Dev 2001; 15:2321-42.

96. Soderhall K, Cerenius L. Role of the prophenoloxidase-activating system in invertebrate immunity. Curr Opin Immunol 1998; 10:23-8.

97. Sorrentino RP, Carton Y, Govind S. Cellular immune response to parasite infection in the Drosophila lymph gland is developmentally regulated. Dev Biol 2002; 1;243(1):65-80.

98. Takihara Y, Hara J. Polycomb-group genes and hematopoiesis. Int J Hematol 2000; 72:165-72.
99. Tepass U, Fessler LI, Aziz A et al. Embryonic origin of hemocytes and their relationship to cell death in Drosophila. Development 1994; 120:1829-37.
100. Thummel CS. Files on steroids—Drosophila metamorphosis and the mechanisms of steroid hormone action. Trends Genet 1996; 12:306-10.
101. Torok I, Herrmann-Horle D, Kiss I et al. Down-regulation of RpS21, a putative translation initiation factor interacting with P40, produces viable minute imagos and larval lethality with overgrown hematopoietic organs and imaginal discs. Mol Cell Biol 1999; 19:2308-21.
102. Torok I, Strand D, Schmitt R et al. The overgrown hematopoietic organs-31 tumor suppressor gene of Drosophila encodes an Importin-like protein accumulating in the nucleus at the onset of mitosis. J Cell Biol 1995; 129:1473-89.
103. Torok T, Harvie PD, Buratovich M et al. The product of proliferation disrupter is concentrated at centromeres and required for mitotic chromosome condensation and cell proliferation in Drosophila. Genes Dev 1997; 11:213-25.
104. Tsang AP, Fujiwara Y, Hom DB et al. Failure of megakaryopoiesis and arrested erythropoiesis in mice lacking the GATA-1 transcriptional cofactor FOG. Genes Dev 1998; 12:1176-88.
105. Van Oostveen J, Bijl J, Raaphorst F et al. The role of homeobox genes in normal hematopoiesis and hematological malignancies. Leukemia 1999; 13:1675-90.
106. Vincent S, Vonesch JL, Giangrande A. glide directs glial fate commitment and cell fate switch between neurones and glia. Development 1996; 122:131-139.
107. Wang Q, Stacy T, Binder M et al. Disruption of the Cbfa2 gene causes necrosis and hemorrhaging in the central nervous system and blocks definitive hematopoiesis. Proc Natl Acad Sci USA 1996; 93:3444-9.
108. Watson KL, Johnson TK, Denell RE. Lethal(1) aberrant immune response mutations leading to melanotic tumor formation in Drosophila melanogaster. Dev Genet 1991; 12:173-87.
109. Yan R, Small S, Desplan C et al. Identification of a Stat gene that functions in Drosophila development. Cell 1996; 84:421-30.
110. Zeidler MP, Perrimon N, Strutt DI. Polarity determination in the Drosophila eye: A novel role for unpaired and JAK/STAT signaling. Genes Dev 1999; 13:1342-53.
111. Ziegler BL, Valtieri M, Porada GA et al. KDR receptor: A key marker defining hematopoietic stem cells. Science 1999; 285:1553-8.

CHAPTER 11

Intraembryonic Development of Hematopoietic Stem Cells during Human Ontogeny:
Expression Analysis

Caroline Marshall

The adult hematopoietic system is composed of a number of different cell types, including erythrocytes and cells of the myeloid and lymphoid lineages. It is generally believed that all these cell types derive, through a series of maturing progenitors, from a common stem cell, which first appears during embryogenesis and persists into adult life. This hematopoietic stem cell (HSC) is defined by its ability to self-renew and to generate cells of all hematopoietic lineages. Clearly these cells would have enormous therapeutic potential in the treatment of blood disorders. However, because of the circulatory nature of the hematopoietic system and the multiple cell types involved, the processes controlling the generation and development of HSCs have proved difficult to study and are poorly understood.

Thus far, putative HSCs have been isolated from bone marrow, peripheral blood, umbilical cord blood and fetal liver largely on the basis of expression of the membrane glycoprotein CD34. In mice this CD34-positive (CD34[+]) population is able to reconstitute all hematopoietic lineages in myeloablated recipients. However, CD34 expression is not restricted to hematopoietic cells, nor is its function in terms of blood cell development established. Moreover, the existence of human hematopoietic cells with similar reconstituting ability but that do not express CD34 has been demonstrated.[1] The identification of more specific and developmentally relevant molecules is required to increase our understanding of how the hematopoietic system is formed, how HSCs are maintained and how to culture these comparatively rare cells ex vivo.

During human embryogenesis, definitive hematopoiesis in the fetal liver commences around the fifth week of gestation and subsequently switches to other hematopoietic tissues including the bone marrow, thymus and spleen. The discovery of cells with pluripotent hematopoietic repopulating ability in a region of the embryo distinct from the extra-embryonic yolk sac and preceding the onset of fetal liver hematopoiesis suggested a previously unknown source of HSCs. These cells were initially identified in the mouse at 8.5 days postcoitum (dpc)[2-4] and subsequently in the 4-6 week human embryo[5] and mapped to a region of the intraembryonic splanchnopleural mesoderm comprising the dorsal aorta, gonadal ridge and mesonephros, termed the AGM region. Indeed, cells with lymphoid and myeloid potential can be detected in the

Hematopoietic Stem Cell Development, edited by Isabelle Godin and Ana Cumano.
©2006 Eurekah.com and Kluwer Academic / Plenum Publishers.

murine splanchnopleura as early as 7.5 dpc, before circulation between the embryo and yolk sac is established.[6]

HSC activity within the embryonic AGM region is tightly regulated both spatially and temporally and precedes the appearance of the various lineage-restricted cells that subsequently develop to form the complete definitive hematopoietic system. It therefore provides a relatively simple environment in which to study gene expression during HSC induction and development prior to and distinct from lineage commitment.

Techniques

There are a number of established techniques available for the analysis of gene/protein expression in fixed tissues and isolated cells. Whole embryos can be embedded in wax or frozen and cut into thin sections to expose cells held within their natural environment. These tissue sections can then be incubated with antibodies raised against a protein of interest. Coupling the antibody to a simple enzyme-substrate detection system allows the expression of the protein to be visualised by immunohistochemistry. Alternatively, where no antibody is available, radioisotope-labelled probes can be used to detect RNA expression by in-situ hybridisation. However, it should be noted that the presence of RNA transcripts within a cell at a given time is not necessarily an indication of the proteins, such as surface growth factor receptors, that are expressed.

The expression levels of a number of proteins can be analysed simultaneously by incubating dissociated living cells with fluorescent-labelled antibodies and comparing the relative fluorescence by flow cytometry. In a similar way, viable cells can be isolated from tissues by a process of fluorescence-activated cell sorting (FACS) for subsequent culture.

Another approach is to generate transgenic animals in which a reporter gene, such as beta-galactosidase, has been inserted into a gene of interest at the embryonic blastocyst stage. Gene expression within individual cells of transgenic animals can be detected and followed during development via expression of the reporter gene. These methods provide powerful tools with which to study the differential expression pattern of growth factors, cell-surface receptors and downstream signalling molecules both temporally and spatially and have been used extensively to identify and characterise potential HSCs within the embryonic AGM region.

The purpose of gene expression analysis within the AGM region is threefold. First, to confirm that cells within the intra-aortic clusters are indeed true hematopoietic stem cells. Second, common expression patterns exhibited by cluster cells and neighbouring cells would provide clues as to their cellular origin. For some time the possibility of a 'hemangioblast' precursor cell from which both hematopoietic and endothelial lineages are generated has been discussed. Whilst cells with such potential have been demonstrated in vitro, the existence of hemangioblasts in vivo has yet to be proven.[7] Finally, the identification of candidate factors contained within the surrounding microenvironment would be an important step forward in our understanding of how HSCs are produced and regulated in hematopoietic tissues and, therefore, our ability to reproduce these conditions in culture. To this end, gene expression analysis in the embryonic AGM has been extensively studied by a number of researchers and the results have provided a fascinating insight into the generation and development of HSCs during human ontogeny.

Analysis of Gene Expression Patterns in the Human AGM

Within the human embryonic AGM region, HSC activity can be detected between 4-6 weeks gestation. It is spatially restricted along the anterior-posterior axis of the embryo to the region between the upper limb bud and umbilical vein. Transverse sections of tissue taken through this preumbilical region reveal clusters of cells which appear to bud out from the ventral wall of the dorsal aorta (Fig. 1). The appearance of these clusters coincides with HSC

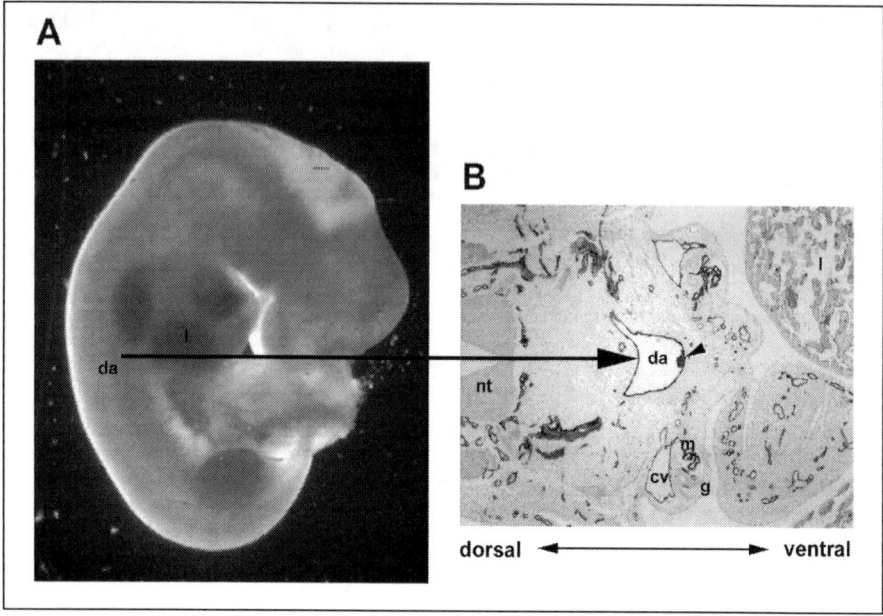

Figure 1. A) 34-day human embryo. Arrow placed at level of preumbilical aorta-gonad-mesonephros (AGM) region. B) Transverse section through embryo at level of arrow immunostained for the membrane glycoprotein CD34 (CD34+ cells indicated by brown colouration). A cluster of CD34-expressing cells can be seen adhering to the ventral wall of the dorsal aorta (arrowhead). da: dorsal aorta; l: liver; nt: neural tube.

activity in the AGM region, between 4-6 of weeks gestation and is restricted to the ventral aspect. In common with other vessels, the wall of the dorsal aorta is lined with a single layer of vascular endothelial cells. Immunohistochemical analysis has shown that both vascular endothelial cells and cells within the associated clusters express CD34[5,8,9] and the platelet-derived endothelial cell adhesion molecule (PECAM)/CD31. However, the two populations can be distinguished from each other by expression of the lectin *Ulex europaeus*, which is restricted to endothelial cells lining the aortic wall.[5] In contrast, the hematopoietic-specific pan-leukocyte marker CD45 is exclusively expressed on cells within the cluster and defines these cells as belonging to the hematopoietic lineage.[5,9] Similar intraluminal cell clusters have been reported associated with the walls of the vitelline artery where it connects with the dorsal aorta.[10] This unique population of CD34+/CD45+ cells within the 4-6 week human embryonic AGM is believed to comprise the preliver HSC compartment and, therefore, the first site of adult-type HSC generation.

Expression of Receptors and Ligands in the Human AGM

It is thought that the decision of a cell to commit to a particular differentiation pathway is largely determined by extrinsic signals that trigger downstream events culminating in the expression of lineage-restricted genes. Generally, these signals take the form of receptor-ligand binding and require the ligand, such as a cytokine or growth factor, to be available outside the cell, and the appropriate receptor to be expressed on the target cell surface.

A number of extracellular factors have been implicated in hematopoietic stem cell regulation including Flt-3 ligand (FL) and stem cell factor (SCF), which are routinely used to extend the proliferative potential of bone marrow-derived CD34[+] cells in culture. Immunohistochemical analysis has shown that cells within the 5-week human intraaortic CD34[+]/CD45[+] clusters express both FL ligand and the corresponding stem cell tyrosine kinase receptor STK-1, human homologue of the murine Flt-3 receptor.[9] Interestingly, endothelial cells lining the wall of the dorsal aorta also express FL and STK-1. However, compared to adjacent endothelium, FL appears to be expressed at a higher level within the cluster cells. Previous studies suggest a role for FL in promoting early progenitor/stem cell proliferation.[11] This difference in ligand expression levels within the AGM may reflect differential signalling intensity related to, or indeed determining, cell type. In contrast, the expression of c-Kit, the receptor for SCF, is restricted to cells within the intraluminal clusters at both RNA and protein level.[8]

As previously stated, putative HSCs within the AGM arise within and are closely associated with vessel walls. During murine embryogenesis, organised vessel formation requires signalling through the tyrosine kinase receptor for vascular endothelial growth factor (VEGF), Flk-1.[12] Expression analysis in the 5-week human embryo reveals that KDR, the human homologue of murine Flk-1, and VEGF ligand are expressed by endothelial cells lining the dorsal aorta.[8,9] At the protein level, KDR and VEGF are also expressed on cells within the CD34[+]/CD45[+] clusters.[9] However, RNA expression of KDR within the cluster appears to be restricted to the basal layer, closest to the endothelium.[8] If hematopoietic cells are indeed budding out from the endothelial layer, this loss of RNA but not protein expression as cells move away from the ventral aortic wall may reflect a downregulation of KDR concomitant with acquisition of a hematopoietic phenotype.

Although *flk-1*-deficient mutant mice have reduced numbers of hematopoietic progenitors, a specific role for KDR/Flk-1 in hematopoiesis has yet to be established. In contrast, mice lacking the endothelial-specific tyrosine kinase receptor Tie-1 have no known hematopoietic defects and this molecule appears to play a role solely in established vessel integrity.[13] The AGM region of murine embryos aged between 9.5-11.5 dpc also bears clusters of CD34+ cells preceding the onset of liver hematopoiesis.[14] Within this region at 10.5 dpc, Tie-1 is expressed on vascular endothelial cells and on cells within the intraaortic clusters.[15] This expression pattern, in the absence of a functional role for Tie-1 in hematopoiesis, suggests that CD45+ cells within the clusters may be generated directly from cells within the endothelial wall. Expression of endothelial markers, such as KDR (Flk-1) and Tie-1 are carried over initially in these cells but subsequently down-regulated.

Expression of Transcription Factors in the AGM

The generation of transgenic mice lacking defined transcription factors has identified these proteins as essential for the development of adult-type/definitive hematopoietic cells during embryogenesis. Amongst these are SCL/tal-1, GATA-2, c-myb and AML1. Disruption of the genes encoding these factors results in a block in all definitive hematopoietic lineages, suggesting that they act, at least initially, at the level of an early stem cell or progenitor.[16-19]

SCL is required for the development of all hematopoietic cells, of both primitive (yolk sac-derived) and definitive lineages. In the 5-week human embryo SCL protein is clearly expressed in both cluster cells and the aortic endothelium whilst SCL RNA expression appears to be restricted to cells within the aortic clusters, an apparent reversal of the pattern for KDR.[8,9]

In mice lacking GATA-2 and c-myb expression, definitive but not primitive hematopoiesis is affected suggesting that these two factors may act directly on committed HSCs/progenitors at a slightly later stage to SCL. Both GATA-2 and c-myb RNA expression is detectable specifically in cells in the aorta-associated clusters in the 5-week human AGM region, but not in neighbouring endothelial cells.[8] GATA-3 RNA transcripts are also detectable in cells within the intraaortic clusters and also in scattered cells underlying the ventral floor of the aorta.[8]

However, these subaortic cells do not appear to express other markers associated with endothelial or hematopoietic cells within the region, such as KDR and SCL, and their identity and possible role in AGM hematopoiesis has yet to be established.

The transcription factor AML1 is also required for normal definitive hematopoiesis. The generation of a mouse model in which normal *AML1* gene expression has been coupled to a reporter gene reveals that within the murine embryonic AGM at 8.5-10.5dpc AML1 is restricted to a subset of endothelial cells in the ventral floor of the dorsal aorta.[20] A limited number of AML1+ cells are also found in the endothelium of the vitelline and umbilical arteries in regions where intraluminal clusters of adherent CD34+ cells have also been reported. At 10.5 dpc, AML1+ cell clusters are found associated with the endothelium of these vessels. Transient AML1 endothelial expression therefore appears to correspond with and precede sites of putative HSC emergence. This is further supported by the absence of intraaortic CD34+ clusters in mice lacking *AML1* gene expression.[20] It is possible that AML1 expression may mark 'hemangioblast'-like cells undergoing transition to the hematopoietic lineage. The expression pattern of AML1 in the human embryo has yet to be investigated.

How the spatial progression from hemangioblast to HSC may be related to changes in expression patterns of molecules described above is illustrated in Figure 2.

Expression of Cell Adhesion Molecules in the AGM

Cell-cell/cell-substrate communication within tissues plays an important role in cell behaviour and fate determination. For example, interactions between blood stem cells/precursors and underlying stromal components regulate cell proliferation and differentiation within the bone marrow. Such interactions within tissues are frequently mediated by cell adhesion molecules (CAMs). CAMs may therefore be involved in generating the clusters of hematopoietic cells associated with the ventral floor of the aorta in the embryonic AGM or indeed in maintaining the integrity of the cluster within the aortic lumen during its formation.

Hematopoietic cell antigen (HCA/ALCAM/CD166), a member of the immunoglobulin family, is expressed at the surface of all the most primitive CD34+ hematopoietic cells in fetal liver and bone marrow, by subsets of nonhematopoietic stromal cells contained within these sites and some other nonhematopoietic embryonic tissues. Within the 4-6 week human AGM, HCA RNA expression around the dorsal aorta is mainly concentrated in the region underlying the ventral wall, leading to the suggestion that HCA may play a role in mediating interactions between HSCs and supporting stromal cells during HSC generation or development.[21]

In contrast to HCA, the hematopoietic cell adhesion molecule (HCAM) is strongly expressed on the surface of cells within the intraaortic clusters in the 5-week human AGM region but absent from underlying and endothelial cells[22] (Marshall unpublished). HCAM is associated with cell migration and homing. Interestingly, expression of Wiskott-Aldrich syndrome protein (WASP), also thought to play an important role in cell homing during immune responses,[23] is similarly restricted to AGM cell clusters at this stage of development (Marshall unpublished). It is possible that these molecules may be involved in the subsequent migration of AGM-derived HSCs, via the circulatory system, to seed the fetal liver.

At 5-weeks of gestation, the vascular-associated adhesion molecules VCAM-1 (CD106) and VE-cadherin are expressed, not surprisingly, by endothelial cells lining the dorsal aorta as well as other embryonic vessels. Both molecules are also expressed by the associated hematopoietic clusters possibly reflecting a common, recent origin (Marshall unpublished).

Common expression patterns shared between endothelial cells lining the dorsal aorta and cells within the hematopoietic clusters in the embryonic AGM are summarised in Table 1.

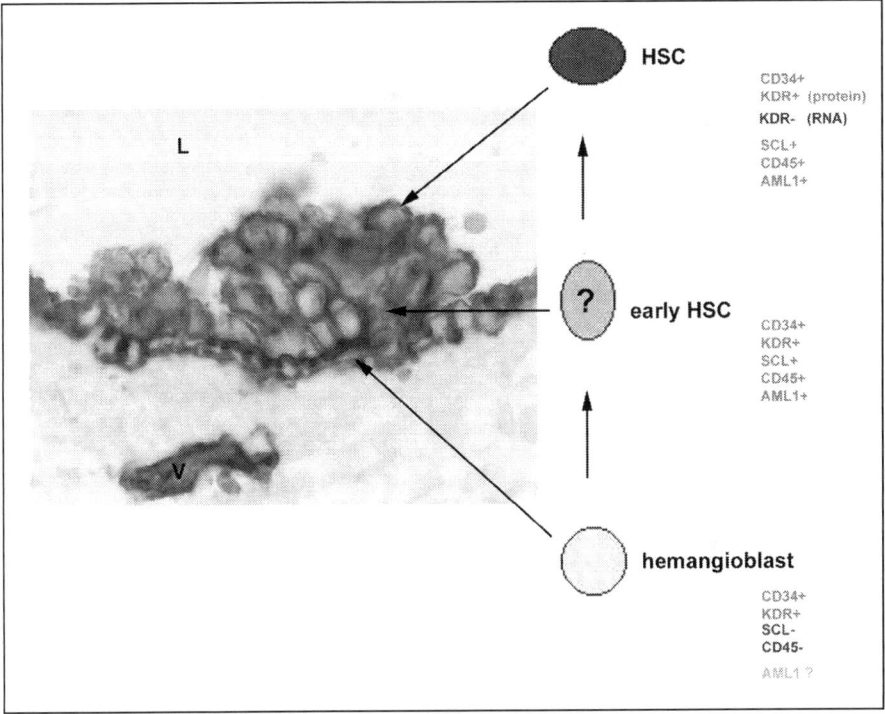

Figure 2. Sequential expression of markers during HSC development in the 34-day human embryonic AGM region. Transverse section through the dorsal aorta within the AGM region immunostained with antibody to CD34. Positive cells are indicated by brown colouration. Progression through putative hemangioblast, early HSC and HSC types in the ventral cell cluster is indicated by arrows. Progressive changes in expression patterns of specific molecules is indicated for each cell type. L: lumen of dorsal aorta. V: transected small blood vessel lined with CD34+ endothelial cells.

Environmental Expression Patterns in the AGM

The ability of a single cell to change its identity during development is well documented. If the cell is mobile, it can migrate through different environments, encountering different extracellular signals, which, if the cell is receptive, can alter its mobility, shape and phenotype. However, if the cell is held within a defined structure, it may only respond to changes in the immediate microenvironment. A powerful example of how this is achieved during embryogenesis is the patterning and shaping of tissues such as the lung by opposing gradients of two or more morphogens and/or antagonists. Cells in the zone of highest concentration of a particular morphogen will respond differently to those at the interface and those furthest away. In this way a 'niche' is created in which the placement of a cell may be critical to its development. An understanding of how the immediate surroundings of a cell change through time and space is necessary to explain the behaviour of that cell.

The identification of a morphological polarity across the dorso-ventral axis of the embryonic dorsal aorta suggested the existence of a localised microenvironment which could regulate cell fate both spatially and temporally and therefore explain the highly restricted appearance of intraaortic hematopoietic clusters.[9] Around the aorta, cells underlying the ventral floor are more closely packed together compared to those adjacent to the lateral and dorsal aspects. The possibility that this cell-dense region could represent a hematopoietic stroma

Table 1. *Comparison of molecules expressed by endothelial cells lining the dorsal aorta and by cells within intraaortic haematopoietic clusters in 4-6 week human embryonic AGM. Shared expression patterns suggest a common haemangioblast precursor.*

	Lineage Association	Intraaortic Hematopoietic Cell	Aortic Endothelial Cell
CD34	H, E	+	+
CD31(PECAM)	E	+	+
CD45	H	+	-
c-kit	H	+	-
Flk-1/KDR	E	+	+
SCL	H	+	+
GATA-2	H	+	+
VE-cadherin	E	+	+
VCAM-1	E	+	+
c-myb	H	+	-
WASP	H	+	-
Lectin	E	-	+

H: hematopoietic lineage; E: endothelial cell

supporting the localised and highly regulated development of HSC has led to an extensive analysis of molecules expressed within it.

As already described the stromal-associated cell adhesion molecule HCA is expressed on the surface of cells corresponding to this region in the 4-6 week human embryonic AGM.[21] Furthermore, smooth muscle α-actin (SMα-A), a marker for human bone marrow stromal cells, is expressed in multiple layers of cells underlying the ventral wall of the dorsal aorta specifically associated with the presence of hematopoietic clusters. Dorsal and lateral aspects of the aortic endothelium, and ventrally outside the AGM, are lined with only a single layer of SMα-A+ cells.[10] Together these data support the presence of stromal cells ventral to the dorsal aorta associated with sites of hematopoietic activity.

An important class of molecules associated with cell proliferation and fate determination during embryogenesis and organogenesis are the extracellular matrix (ECM) molecules. These molecules function extracellularly to influence cell shape, physiology and motility and may also play a role in creating or maintaining morphogen gradients. Tenascin, an ECM molecule closely associated with the transformation of cells at the epithelial-mesenchymal interface in a number of structures during embryonic development, is also upregulated at sites of abnormal cellular activity such as wound healing and tumour formation in established tissues. The ECM molecule fibronectin is also expressed during these processes at higher than basal levels.

During embryogenesis, at 5 weeks of gestation, fibronectin is expressed uniformly around the entire aorta suggesting that it may play a role in vessel formation.[9] Tenascin, in contrast, exhibits a very distinctive and highly regulated pattern of expression and is concentrated in a tight 'knot' within the stromal-like region underlying the ventral wall of the dorsal aorta.[9] This expression is specifically associated with the presence of hematopoietic clusters at the luminal surface since, as the cluster may be skewed to right or left descending the aorta, the 'knot' of tenascin is invariably coincident. Conversely in more caudal regions, where no hematopoietic clusters are observed, the level of tenascin expression appears lower within the stromal-like region than in the surrounding mesenchyme. The precise role of tenascin in embryonic hematopoiesis has yet to be investigated, however, this distinctive expression pattern suggests that it may be involved in the generation of HSCs within the AGM region, if not directly then possibly by recruiting and/or stabilising appropriate hematopoiesis-inducing factors.

One group of factors which are known to play a critical role in cell fate determination throughout embryonic development are the transforming growth factor-β (TGF-β) superfamily of secreted polypeptide growth factors. This family includes TGF-β1 and bone morphogenetic protein-4 (BMP-4) which have been implicated in hematopoietic specification and development in a number of species. BMP-4 has been shown to specify blood formation, via the induction of hematopoietic transcription factors, in *Xenopus* embryos.[24] TGF-β1 can inhibit the initial proliferation and differentiation of long term repopulating-HSCs in culture.[25] At 4-6 weeks gestation, BMP-4 is expressed at high levels, and with striking polarity, in the ventral stromal-like region underlying the intraaortic hematopoietic clusters within the human AGM, compared with surrounding tissue.[26] In younger or older embryos, this polarised expression is not detectable (Fig. 3). The related growth factor TGF-β1 is expressed predominantly by hematopoietic cells within the clusters and is scarcely detectable in the underlying 'stromal' region.[26] These contrasting patterns suggest that both BMP-4 and TGF-β1 may be involved in hematopoietic development in the AGM but at different stages. The ventral polarisation of BMP-4 expression suggests an early role in the specification of hematopoietic cells from hemangioblast precursors or mesodermal cells located within the ventral wall of the dorsal aorta. TGF-β1, on the other hand, appears to act at a later stage on the emerging hematopoietic cells, possibly to inhibit proliferation and/or differentiation of nascent HSC prior to their migration to the fetal liver.

Comparison of AGM and Fetal Liver Expression Patterns

Unlike the comparatively simple environment of the dorsal aorta, the liver is composed of multiple cell types, which are organised into a complex network of ducts and canaliculi. Throughout the liver, blood is carried along sinusoids, cavities lined with a monolayer of flattened endothelial cells. Vessels carrying blood to the liver are derived from the vitelline arteries that branch from the dorsal aorta. Theoretically, this could provide a simple route via which intraaortic HSCs circulate from the AGM to seed the fetal liver. Alternatively, HSCs could arise de novo within the liver from preexisting cells dependent on the presence of a supportive microenvironment such as exists in the AGM and bone marrow.

The first CD34+ hematopoietic cells are detected in the fetal liver at around 30-32 days of gestation, after their appearance in the dorsal aorta.[10] The liver continues as the major fetal blood-forming organ up to 16 weeks when hematopoiesis switches to the bone marrow. At 30 days, flattened endothelial cells lining the cavities or capillaries are faintly positive for CD34. Within these spaces, often associated with the endothelium, are round CD34+ cells that coexpress CD45 indicative of a hematopoietic phenotype.[10] Initially these hematopoietic cells are rare and scattered but progressively increase in number reflecting an increase in liver hematopoiesis.

A comparison of CD34+/CD45+ hematopoietic cells FACS sorted from 5-week fetal liver and dorsal aorta by reverse transcriptase-polymerase chain reaction (RT-PCR) analysis of RNA expression shows that the hematopoietic-associated transcription factors GATA-2, GATA-3 and c-myb are expressed by aortic but not liver-derived cells. SCL expression can be detected in both populations at this stage.[8]

The liver is a site of HSC renewal, lineage commitment and differentiation and therefore contains a highly heterogeneous population of hematopoietic cells at different stages of maturation. In this environment stem cells and early progenitors can be distinguished from

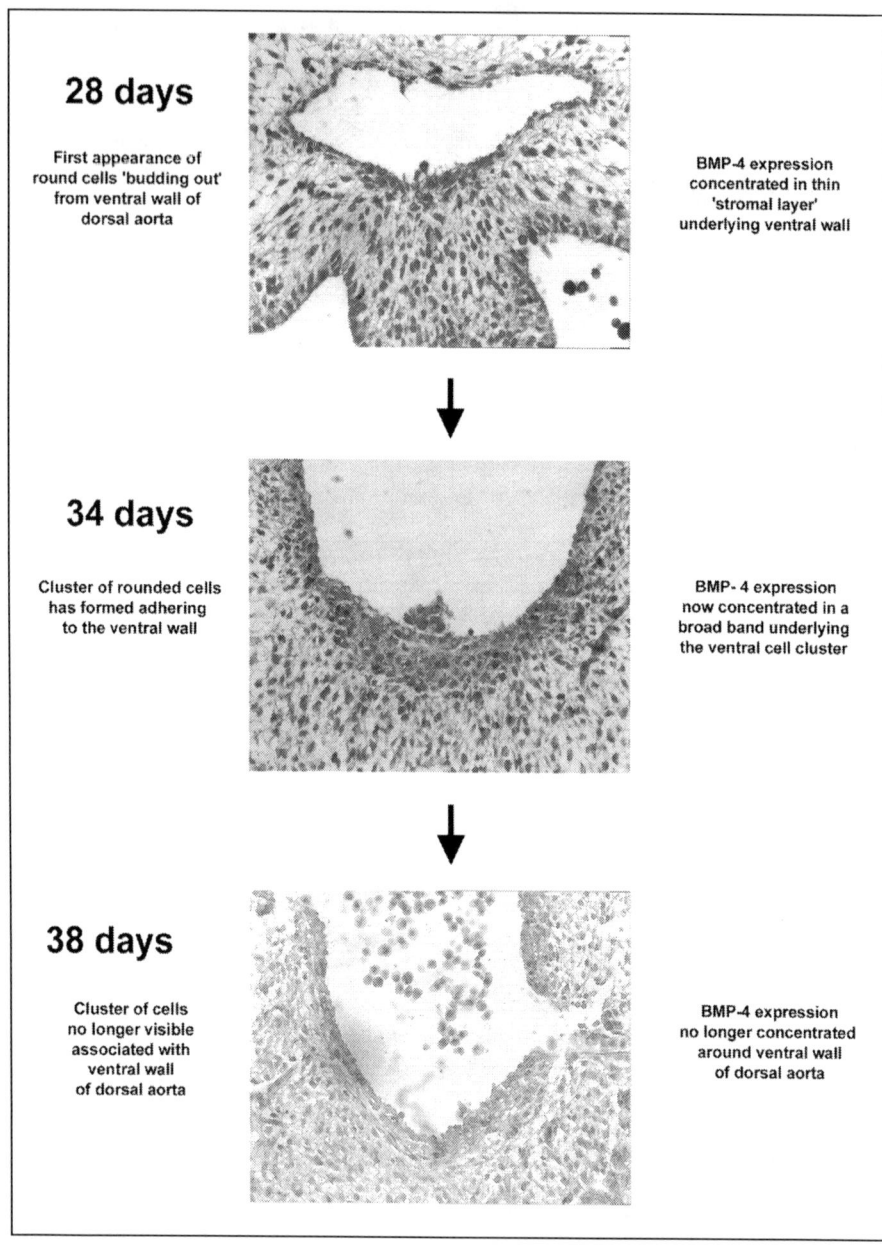

Figure 3. Temporal relationship between hematopoietic cell cluster development and BMP- 4 expression within the human embryonic AGM region. Transverse sections through the AGM regions of human embryos aged 28, 34 and 38 days immunostained with antibody to BMP-4. BMP-4 expressing cells are colored in brown.

more mature cell types on the basis of CD38 expression. CD38 is a membrane glycoprotein expressed on precursors of all hematopoietic lineages but not on pluripotent stem cells.

At 6.5 weeks CD34+/CD38- cells FACS sorted from the fetal liver continue to express SCL RNA and an increase in GATA-2 and -3 transcripts can be detected.[8] By 12 weeks the CD34+/CD38- stem cell/progenitor population also expresses c-myb, c-kit and, at comparatively low levels, KDR. Interestingly transcripts for all these molecules are also detected in the more mature CD34+/CD38+ population at 12 weeks, suggesting that in the liver expression is not restricted to stem cells. Transcripts for AML-1β, FL, STK-1 (flt-3), TGF-β and a number of cytokine receptors have also been detected in CD34+ liver cells at this stage of development.[27]

Summary

Expression analysis within regions of early hematopoietic stem cell activity, particularly the embryonic AGM, has revealed fundamental similarities across species supporting the use of animal models to further our understanding of early human blood cell development.

It has revealed that CD34+/CD45+ cells in the intraaortic clusters within the 4-6 week human embryonic AGM region, the first identified intraembryonic site of HSC activity:

- express factors associated with or essential for hematopoietic stem cell development,
- express a number of markers in common with vascular endothelial cells with which they are in close contact,
- are contained within a microenvironment in which specific factors are expressed at higher levels compared to the surrounding mesenchyme and
- express a number of factors in common with the CD34+/CD38- HSC/early progenitor compartment in the 6-12 week fetal liver.

Together these observations support the stem cell/early progenitor identity of hematopoietic cell clusters within the 4-6 week human embryonic AGM and the existence of a bipotential cell which can give rise to cells of either the hematopoietic or the endothelial lineage. The mechanism by which this is achieved is as yet unknown but is likely to involve the switching of a genetic programme within an immature cell, followed by the proliferation of that cell, in response to localised extracellular factors.

This approach has also emphasised that it may be more accurate to define hematopoietic stem cells, and possibly all stem cells, not only by the molecules they themselves express but also by factors expressed within the microenvironment, or niche, in which they reside.

It is likely that the progressive commitment from mesodermal to hematopoietic stem cell within the embryonic AGM, as well as other hematopoietic tissues, involves a complex collaboration of factors. Whilst the presence of some candidate factors, such as BMP-4 and TGF-β has been demonstrated, a far more extensive analysis of other morphogen families is required. Recent technological developments such as genechip expression analysis may allow us to identify precise molecular signalling pathways involved in commitment to the hematopoietic lineage and pinpoint critical extracellular factors. An understanding of the first stages of HSC induction and regulation would greatly improve our ability to expand and maintain HSC in culture and, consequently, their usefulness as therapeutic tools.

The enormous importance of stem cell biology is now widely acknowledged. In recent years there have been major advances in the understanding of these remarkable cells in a number of systems, opening up the possibility of their use in the repair of adult tissues. The hematopoietic system presents a particular challenge because of the number of progenitor and mature cell types generated and the complexity of the microenvironment in which HSCs reside. Uncovering the secrets of hematopoietic stem cells will not only be a major scientific and clinical breakthrough but may also teach us valuable lessons about stem cell commitment and maturation in general.

References

1. Bhatia M, Bonnet D, Murdoch B et al. A newly discovered class of human hematopoietic cells with SCID- repopulating activity. Nat Med 1998; 4(9):1038-1045.
2. Medvinsky AL, Samoylina NL, Muller AM et al. An early preliver intraembryonic source of CFU-S in the developing mouse. Nature 1993; 364(6432):64-67.
3. Medvinsky A, Dzierzak E. Definitive hematopoiesis is autonomously initiated by the AGM region. Cell 1996; 86(6):897-906.
4. Godin I, Dieterlen-Lievre F, Cumano A. Emergence of multipotent hemopoietic cells in the yolk sac and paraaortic splanchnopleura in mouse embryos, beginning at 8.5 days postcoitus. Proc Natl Acad Sci USA 1995; 92(3):773-777.
5. Tavian M, Coulombel L, Luton D et al. Aorta-associated CD34+ hematopoietic cells in the early human embryo. Blood 1996; 87(1):67-72.
6. Cumano A, Dieterlen-Lievre F, Godin I. Lymphoid potential, probed before circulation in mouse, is restricted to caudal intraembryonic splanchnopleura. Cell 1996; 86(6):907-916.
7. Choi K, Kennedy M, Kazarov A et al. A common precursor for hematopoietic and endothelial cells. Development 1998; 125(4):725-732.
8. Labastie MC, Cortes F, Romeo PH et al. Molecular identity of hematopoietic precursor cells emerging in the human embryo. Blood 1998; 92(10):3624-3635.
9. Marshall CJ, Moore RL, Thorogood P et al. Detailed characterization of the human aorta-gonad-mesonephros region reveals morphological polarity resembling a hematopoietic stromal layer. Dev Dyn 1999; 215(2):139-147.
10. Tavian M, Hallais MF, Peault B. Emergence of intraembryonic hematopoietic precursors in the preliver human embryo. Development 1999; 126(4):793-803.
11. Small D, Levenstein M, Kim E et al. STK-1, the human homolog of Flk-2/Flt-3, is selectively expressed in CD34+ human bone marrow cells and is involved in the proliferation of early progenitor/stem cells. Proc Natl Acad Sci USA 1994; 91(2):459-463.
12. Shalaby F, Rossant J, Yamaguchi TP et al. Failure of blood-island formation and vasculogenesis in Flk-1-deficient mice. Nature 1995; 376(6535):62-66.
13. Sato TN, Tozawa Y, Deutsch U et al. Distinct roles of the receptor tyrosine kinases Tie-1 and Tie-2 in blood vessel formation. Nature 1995; 376(6535):70-74.
14. Wood HB, May G, Healy L et al. CD34 expression patterns during early mouse development are related to modes of blood vessel formation and reveal additional sites of hematopoiesis. Blood 1997; 90(6):2300-2311.
15. Marshall CJ, Thrasher AJ. The embryonic origins of human haematopoiesis. Br J Haematol 2001; 112(4):838-850.
16. Shivdasani RA, Mayer EL, Orkin SH. Absence of blood formation in mice lacking the T-cell leukaemia oncoprotein tal-1/SCL. Nature 1995; 373(6513):432-434.
17. Tsai FY, Keller G, Kuo FC et al. An early haematopoietic defect in mice lacking the transcription factor GATA-2. Nature 1994; 371(6494):221-226.
18. Mucenski ML, McLain K, Kier AB et al. A functional c-myb gene is required for normal murine fetal hepatic hematopoiesis. Cell 1991; 65(4):677-689.
19. Okuda T, van Deursen J, Hiebert SW et al. AML1, the target of multiple chromosomal translocations in human leukemia, is essential for normal fetal liver hematopoiesis. Cell 1996; 84(2):321-330.
20. North T, Gu TL, Stacy T et al. Cbfa2 is required for the formation of intra-aortic hematopoietic clusters. Development 1999; 126(11):2563-2575.
21. Cortes F, Deschaseaux F, Uchida N et al. HCA, an immunoglobulin-like adhesion molecule present on the earliest human hematopoietic precursor cells, is also expressed by stromal cells in blood-forming tissues. Blood 1999; 93(3):826-837.
22. Watt SM, Butler LH, Tavian M et al. Functionally defined CD164 epitopes are expressed on CD34(+) cells throughout ontogeny but display distinct distribution patterns in adult hematopoietic and nonhematopoietic tissues. Blood 2000; 95(10):3113-3124.
23. Thrasher AJ, Jones GE, Kinnon C et al. Is Wiskott—Aldrich syndrome a cell trafficking disorder? Immunol Today 1998; 19(12):537-539.

24. Maeno M, Mead PE, Kelley C et al. The role of BMP-4 and GATA-2 in the induction and differentiation of hematopoietic mesoderm in Xenopus laevis. Blood 1996; 88(6):1965-1972.

25. Sitnicka E, Ruscetti FW, Priestley GV et al. Transforming growth factor beta 1 directly and reversibly inhibits the initial cell divisions of long-term repopulating hematopoietic stem cells. Blood 1996; 88(1):82-88.

26 Marshall CJ, Kinnon C, Thrasher AJ. Polarized expression of bone morphogenetic protein-4 in the human aorta-gonad-mesonephros region. Blood. 15;96(4):1591-3.

27. Oh IH, Lau A, Eaves CJ. During ontogeny primitive (CD34(+)CD38(-)) hematopoietic cells show altered expression of a subset of genes associated with early cytokine and differentiation responses of their adult counterparts. Blood 2000; 96(13):4160-4168.

CHAPTER 12

Gene and Cell Therapy Involving Hematopoietic Stem Cell

Isabelle André-Schmutz and Marina Cavazzana-Calvo

Abstract

Hematopoietic Stem Cells (HSC) possess two characteristics, long term self-renewal capacity and pluripotentiality, that render them able to develop into the different blood cell lineages. HSC transplantations (HSCT) have thus been used to reconstitute hematopoiesis after myeloablation for more than three decades. The range of therapeutic applications of HSCT has increased from hematological malignancies to immune deficiencies, solid tumors and auto-immune diseases. Until recently, HSC were collected from bone marrow. Additional sources of HSC can be used, like peripheral blood from cytokine-mobilized donors and cord blood. Although autologous HSCT remains the most frequent approach, mainly for malignant diseases, allogeneic HLA compatible or partially HLA incompatible HSCT are more and more frequently used. This has been made possible thanks to the development of techniques of HSC selection. The present chapter will review the different possibilities of HSC sources and donors, the various indications of HSCT used either for cell or gene therapy.

Hematopoietic Stem Cell Transplantation (HSCT): Generalities

Autologous and allogeneic hematopoietic stem cell transplantation (HSCT) is the first achievement in the field of cellular therapy. Based on the research carried out by Van Bekkun and Billingham, HSC allogeneic grafts were first performed at the end of the 1960s and aimed to replace the recipient's "diseased" lymphohematopoietic system with a healthy system from an individual of the same species. From the time the first clinical studies were carried out (1967), the approach developed rapidly. The number of patients treated with HSCT in Europe increased from several hundreds per year in the eighties to more than 6.000 in 2000, with some degree of variation in the distribution of the type of donor and stem cell source. Today, stem cells from bone marrow, cytokine-mobilised peripheral blood and cord blood from a family donor or an unrelated matched donor can be used. The development of international registries of volunteer donors explains, at least to some extent, the two-fold increase in the number of grafts performed over the last 10 years. Although the principal indication for this therapy remains malignant hemopathies (85%), other diseases, such as inherited immunodeficiencies, metabolic diseases and red cell deficiencies are currently treated by allogeneic HSCT.[1,2]

In Europe, the increase in the number of autologous HSCT performed is even higher; in 2000 more than 12,500 patients underwent autologous grafts, 80% for malignant hemopathy and 18% for solid tumours.

Hematopoietic Stem Cell Development, edited by Isabelle Godin and Ana Cumano.
©2006 Eurekah.com and Kluwer Academic / Plenum Publishers.

HSCT requires the use of a hematopoietic cell that is sufficiently immature to repopulate the bone marrow of an individual, at least partially, and to ensure peripheral blood homeostasis throughout life. The ex vivo processing of HSC has three objectives: firstly, to reduce the effects of chemo- and/or radiotherapy-induced cytopenia; secondly, and this is more theoretical than scientific, to reduce tumoral contamination of certain autologous grafts. Furthermore, allogeneic HSCT, unlike autologous HSCT, has a therapeutic purpose related to the immunological activities of the mature T lymphocytes contaminating it. Correction of chemo and/or radiotherapy-induced aplasia before HSCT raises several issues: the minimum number of hematopoietic progenitor cells to inject, the hematopoietic quality of any cellular product intended for grafting, and the more recent issue regarding in vitro expansion/differentiation into one or more cell lines.

Donors of Hematopoietic Stem Cells and Impact of the Degree of HLA Compatibility

Although the majority of allogeneic bone marrow transplantation have been performed using HLA-identical sibling or closely matched family donor (62% in 2000), alternative donor sources such as unrelated (31%) or more highly HLA-disparate family donors (7%) have emerged as viable alternatives.[2] Many different studies have shown an inverse correlation between HLA disparity between the donor and the recipient and the rate of graft rejection and Graft-versus-Host Disease.[3-5] HLA disparity can be classified according to the techniques required to reveal them in antigen mismatches (revealed by serological techniques) and allele mismatches (determined by DNA typing or sequencing). Two meticulous studies were performed on more than 400 patients grafted with bone marrow stem cells from matched-unrelated donors for malignant hemopathies to determine the impact of antigen or allele mismatches on the clinical outcome of HSCT.[6,7] The first showed that most patients who received a transplant from donors compatible for HLA A, B and DR antigens as determined by serological techniques were mismatched at the allelic level. In addition, in the second study, Petersdorf et al demonstrated that class I antigens mismatches induced a statistically significant higher rate of graft rejection than class II allele mismatches.[6] The explanation for the difference in rates of rejection may rely on the fact that antigen mismatches implicates amino-acids substitution in regions of contact with both HLA molecules and T cell receptor, while allele mismatches concern less substitutions only in the region of contact with HLA molecules.

As discussed below, T-cell depletion of the graft and intensification of the conditioning regimen have demonstrated a good efficiency in prevention of these complications when partially incompatible donors are used. A recent study comparing T cell depleted HSCT from phenotypically matched unrelated donors, one antigen mismatched unrelated donors and haploidentical related donors demonstrated identical rates of engraftment and incidence of acute or chronic GVHD, but higher incidence of relapse in the haploidentical group.[8] The addition of Thymoglobulin in the conditioning regimen of the last group could explain both the low rate of GVHD and the higher incidence of relapse. Transplant Related Mortality was higher in the two HLA disparate groups than in the phenoidentical group. Overall survival was significantly better (58%) for the group of matched unrelated donor than for the two other groups (34 and 21%). Another retrospective analysis performed on children grafted with geno-identical or pheno-identical group with 0, 1 or 2 HLA class I mismatches demonstrated a similar survival and rate of engraftment between geno and matched unrelated donors, while the presence of one or more mismatch was deleterious on both criteria.[9,10]

At present, most centers employ phenotypically matched HSCT as the next best option for patients who do not have an identical or closely matched related donor. In case of absence of any suitable matched unrelated donor (MUD) or when HSCT has to been performed within a time period that does not allow finding such donor because of the severity or fast progression of the disease, a more HLA disparate related donor is considered.

Related HSCT with donors (usually the parents) who are mismatched at either 1 or 3 HLA antigens, i.e., haploidentical donors, are employed as next best option for those patients who do not have a geno- or pheno-identical donor. Depending on the degree of HLA incompatibility, a stronger conditioning regimen and immunosuppression as well as T cell depletion of the graft are required to avoid graft rejection and graft versus host disease. When these conditions are fulfilled, transplantation with HSC from haploidentical donors was shown to be quite safe.[8,11-16] Nevertheless, beside their role in GVHD, T lymphocytes play a key role in the protection against infections and relapses.[17,18] Their depletion associated with a delayed T cell reconstitution in the haploidentical setting as compared to MUD and geno-identical HSCT resulted in a higher incidence of severe infections and relapses. Consequently, overall survival was globally lower than for MUD and geno-identical HSCT.[8,10,12,14,16]

To conclude, although the use of partially matched related donors is advantageous by creating immediate donor availability, unrelated pheno-identical donors are preferred overall, when 10 out of 10 HLA antigens are identical.

Sources of Hematopoietic Stem Cells

During thirty years, the major source of HSC has been the bone marrow. Since the early 1990's, peripheral blood stem cells (PBSC) are increasingly used in place of bone marrow cells both for autologous and allogeneic transplantation. They represented more than 80% of the transplantations performed in Europe in 2000.[2] HSC are mobilized in the blood by treatment with granulocyte-colony-stimulating factor (G-CSF). Because of the higher number of both CD34+ and CD3+ cells—more than 1 log difference for both—in the apheresis from mobilized donor, it was hypothesized that transplantation of PBSC would result in higher engraftment, but although higher rate of GVHD. Several studies performed on HLA-identical siblings donors have shown a faster engraftment using PBSC, but conflicting data concerning GVHD issue. Four suggested identical rates of acute and chronic GVHD,[19-22] whereas 3 others described higher rates of cGVHD but similar rates of aGVHD when PBSC were compared to bone marrow stem cells (BMSC).[23-26] Three factors may explain the discrepancy in the rates of cGVHD observed in the different studies: (1) the small number of patients included in most of them, (2) the difference in the prophylaxis against GVHD and, (3) the dose of G-CSF used in the different protocols.[22] In the EBMT/IBMTR analysis performed on several hundreds of patients, the difference in cGVHD incidence was confirmed,[27] but had no impact on overall survival rate, which was similar in both groups. The difference observed in the speed of engraftment may rely on G-CSF used to mobilize HSC in the blood.[26,28-30] Indeed, this treatment results in a higher number of hematopoietic precursors infused when the PB is used compared to normal BM. This hypothesis was confirmed by the absence of difference in speed of engraftment between BMSC and PBSC, when BMSC donors were treated with G-CSF like PBSC donors.

More recently, mobilization of HSC was also performed on matched unrelated donors.[31-33] Except a faster recovery of granulocytes and platelets, no statistically significant difference was noted in the overall clinical outcome of PBSC vs. BMSC transplantation. However, for advanced leukemias, transplanted related mortality and disease free survival were significantly better for patients transplanted with PBSC than with BMSC.[21,22,34,35]

All these data suggest that PBSC is rather safe, despite a higher risk of cGVHD for HLA-identical HSCT, and constitutes a viable alternative source of HSC.

A third source of HSC, umbilical-cord blood (CB), has been investigated since the early 1990's. This last HSC source presents particular biological characteristics due to its ontogenic origin; its self-renewal and proliferation capacities are superior to those of adult bone marrow cells,[36-38] allowing them to be ideal target cells when available to correct inherited monogenic diseases by gene transfer.

GVHD incidence was lower than the one expected, even with HLA-mismatched CB.[39-43] This observation may be related to reduced number or immaturity of T lymphocytes in cord blood as compared to BM or PB. Two studies showed no difference in survival between pediatric patients who had received transplants of 0 to 3 HLA antigens mismatched unrelated cord blood or matched unrelated bone marrow.[44,45] CB was then considered as a viable source of HSC and an acceptable alternative to haploidentical related donors, because of the apparent less stringent requirements for HLA-identity between donor and recipient. In addition, CB presents the advantages of rapid availability and very low rate of contamination with herpes group viruses.

Conversely, the low number of HSC in the CB is probably responsible for the slow engraftment and high rejection rate when these cells are used to transplant adults,[46,47] although HLA incompatibility may partially explain this observation. It thus justifies the efforts aimed at expanding in vitro CB HSC without loosing their transplantation capacity.[36-38,48]

Hematopoietic Stem Cell Processing

HSC Selection and Doses

Even in the case of HLA-genoidentity between donor and recipient, unmanipulated bone marrow transplantation results in high incidence of graft-versus-host disease (GVHD). This complication, which can be classified in acute or chronic GVHD on timing of onset and clinical features, remains one of the major cause of morbidity and mortality after allogeneic HSCT. The key role of T cells in the physiopathology of GVHD is known since the 1970s.[49-51] The use of HLA-non genetically identical donors, and more recently for pediatric patients, HLA-haploidentical donors -one of the parent who shares only one HLA haplotype with the child- has been made possible thanks to the development of tools (anti-T monoclonal antibodies) and technologies able to remove T cells.

Other cells in the bone marrow harvest are considered as unwanted in certain circumstances: B lymphocytes which can carry viruses such as Epstein-Barr virus able to induce lymphomas in immunocompromised hosts, and residual tumor cells in the case of autologous graft.

HSC can be purified by positive or negative selection thanks to monoclonal antibodies against differentiation antigens expressed on cell membranes. Negative selection involves monoclonal antibodies used alone.[52] In particular, CAMPATH-1M antibodies directed against CDw52 expressed both on lymphocytes and monocytes[53] combined with rosetting with sheep red blood cells[54] have been extensively used. Positive selection of HSC, which has been developed more recently, is based on the expression of the CD34 antigen by human HSC.[55,56]

The threshold of CD34+ cells required for engraftment in the autologous setting is 3 x 10^6/kg of the recipient's body weight. For allogeneic HSCT, this dose has to be increased depending on the major histocompatibility complex (HLA) disparity between the donor and the recipient. For HLA-identical allogeneic HSCT, the optimal dose of CD34+ cells was recently discussed by Zaucha et al, for whom high doses of CD34+ cells (> 8 x 10^6 CD34+ cells/kg) were associated with accelerated engraftment, but also increased incidence of cGVHD.[57] When an allogeneic graft is only partially compatible, a threshold of 10 x 10^6 CD34+cells/kg, referred to as a HSC megadose, has been suggested.[15] It is also based on data from mice showing that large numbers of immature hematopoietic precursor cells are able to cross the residual immunological barrier of a lethally or sub-lethally irradiated recipient.[58,59] This procedure, carried out extensively because of the availability of clinical grade monoclonal antibodies, has two major drawbacks: insufficient knowledge of the phenotype characteristics of HSC and the loss during positive selection of the accessory cells that could facilitate hematopoietic engraftment. Several recent research articles suggest that CD34 negative stem cells and stromal cells favor engraftment.[60-63] It would therefore appear more appropriate to try to develop biotechnological tools which aim to eliminate those mature cell populations whose functions are well known, rather than encourage methods that select cells that are poorly defined at present. Negative selections of mature T- and B-lymphocytes, CD14[+] monocytes and/or natural CD56[+] killer cells may help to select grafts better adapted to new biological findings.

HSC Ex Vivo Culture

Purified CD34[+] cells can be easily made to differentiate rapidly into more mature progenitor cells or used to transfer a "new" gene. The optimization of so called expansion protocols must take into account "ontogenic" constraints since it has been demonstrated that the potential for self renewal of HSC from CB, BM and PB are different.[36,64]

The recent cloning of cytokines involved in HSC proliferation and differentiation has opened up new possibilities of treatment strategies. All cytokines may be divided more or less into two groups: those involved in the proliferation of immature stem cells, and those involved in their terminal differentiation. Stem Cell factor (SCF), FLT3-L and thrombopoietin (TPO) are the most important cytokines in the first group,[65-68] whereas G-CSF, GM-CSF, M-CSF and erythropoietin are the cytokines involved in the various differentiation routes.[69,70] This subgrouping however is still rather vague since the activity of these cytokines differs greatly depending on the target cell; for example, on the one hand, thrombopoietin stimulates the proliferation and survival of the more immature HSC (CD34[+], CD38[-]),[71] and on the other hand, it acts as a differentiation factor on megakaryocytes producing platelets.[72] Thrombopoietin is also referred to as Megakaryocyte Growth and Differentiation Factor (MGDF). Similarly, G-CSF (Granulocyte Colony-Stimulating Factor) used extensively in the clinical protocols to mobilize HSC is also a powerful inducer of granulocyte differentiation.

HSC can be grown in the presence of various combinations of cytokines and hematopoietic growth factors to increase significantly the number of precursor cells and/or mature cells, with the aim of shortening as much as possible the 10-12 day cytopenic aplasia following conditioning regimen for autologous bone marrow grafting.[73] The disappearance of myeloablative post-chemotherapy neutropenia after the injection of autologous cells cultured for 10 days in the presence of SCF, G-CSF and MGDF was the most significant clinical result reported in the litterature.[74] Other cytokine combinations have been assessed in preclinical models aimed at speeding up platelet reconstitution. Clinical research has been carried out on the SCF, MGDF, FLT3-L and IL-3 combination, which appeared to produce the best results in monkeys.[75] In this study, ex vivo different expanded HSC had normal biological function. More recently, activation of Notch signalling in human CB HSC in vitro was shown to enhance myeloid and lymphoid marrow-repopulating ability in a xenogenic murine model of transplantation.[48]

This technology is referred to, inappropriately, as "in vitro expansion" of HSC, since most culture conditions favor differentiation over expansion and result in net HSC losses. Until recently, there was no available method to increase the number of stem cells present at the start of cultures. In addition, from several studies has emerged a new group of intrinsic factors that could allow expansion without or with a low rate of differentiation of HSC.[76,77] In particular, a member of the Hox family of transcription factor, HoxB4 in the mouse and its human homologue HoxC4, were shown to induce ex vivo expansion of HSC without loss of multipotency and/or engraftment capacity.[78,79]

Other Indications for HSCT

HSCT is associated with a complete renewal of the immune system, an effect that lead some investigators to test it for some indications different than leukemia and hematological deficiencies, mainly solid tumors and autoimmune disorders. Beside the beneficial effect of

HSCT for aplasia induced by high dose chemotherapy in case of solid tumors, the idea for both indications is that resetting the immune system and eradicating aberrant immune cells will in addition allow the development of anti-tumor immune responses and treat severe but limited visceral injuries in case of autoimmune diseases.

In 2000, around 2400 patients with solid tumors were treated with high dose chemotherapy followed by autologous HSCT.[2] The most frequent solid tumors considered as an indication for HSCT are metastatic breast cancer, germinal tumors, neuroblastoma and Ewing sarcoma.[2] Despite a high number of clinical studies, it remains difficult to draw any conclusion about the impact of this approach, mainly because of the small size and heterogeneity of the populations analyzed and the absence of randomization against standard therapy.[80] The case of breast cancer, for which HSCT is associated with high dose chemotherapy as adjuvant therapy, perfectly illustrates this situation. Two phase II clinical studies, which compared included patients to historical groups concluded to a benefitial effect of HSCT.[81,82] Conversely, no advantages were observed by 3 other groups, who performed two-arms trials, comparing autologous HSCT with high dose chemotherapy with either conventional chemotherapy, or autologous T cell depleted or not depleted HSCT. No differences were noted between each group of patients, and especially no difference in disease-free survival or overall survival.[83-85] As underlined by Lippman,[86] the two first trials presented a bias due to the detection of disease stage which was performed much more carefully in the treated group as compared to the historical groups, leading to the inclusion of patients with small metastases, no intercurrent illness and no brain metastases.

The proof of potential role of HSCT for the treatment of autoimmune disorders was given by a few patients who had in addition to their autoimmune disease a life-threatening hematological disorder (aplastic anemia, leukemia), and were then treated by allogeneic HSCT.[1] HSCT has been described as a possible treatment for patients with severe autoimmune diseases such as systemic sclerosis (SS), rheumatoid arthritis (RA), multiple sclerosis (MS) and systemic lupus erythematous (SLE) refractory to conventional treatments.[87-92] Patients with less frequent autoimmune diseases have been included in pilot studies.[2] In particular, some promising results have been obtained in patients suffering from severe juvenile chronic arthritis treated with autologous HSCT.[93]

Although autologous HSCT was shown to lead to a higher rate of relapses than allogeneic HSCT, it is associated with lower transplant-related mortality. Current protocols thus include a conditioning that insures the complete ablation of host lymphohemopoiesis and infusion of $CD34^+$ selected HSCT. Indeed, T cell purging is essential to avoid relapse.[94] Over MS, SLE and SS, which are the most common diseases treated by HSCT, efficacy of HSCT was found to be the best for SLE and SS.[1,95] In 2000, 400 patients were treated by autologous or allogeneic HSCT worldwide in phase I or II clinical trials.

Autoimmune diseases treated by HLA matched siblings also include hemolytic anemia,[96] pure red cell aplasia,[97] Evans syndrome,[98,99] and autoimmune thyroiditis.[100] Non myeloablative or reduced intensity allogenic HSCT protocols are being written for all these indications.

Until recently, the use of HSC to treat inherited diseases other than those of the lymphohematopoietic system was inconceivable. One recent clinical trial has in part modified this view. Shapiro et al have shown that a rapidly progressive neurodegenerative disorder that affects central nervous system myelin as well as adrenal cortex can benefit from HSCT overall when the procedure is performed at an early stage of the disease.[101] X-linked adrenoleukodystrophy is a demyelinating disorder of the central nervous system leading to a vegetative state and death within 3-5 years once clinical symptoms are detectable. The rationale for bone marrow transplantation has relied on the hypothesis that functional bone-marrow cells from the donor could cross the blood-brain-barrier in the recipient and exert a favorable effect on the mechanisms leading to demyelination. The clinical results reported that in eighteen transplanted children indeed provide a proof of this concept, since twelve of them survived. Moreover, CNS disease progression has been halted and a good quality of life maintained. In two out of the twelve long-term surviving children, a complete disappearance of the cerebral lesions was observed on MRI.

Gene Therapy

Gene therapy is a powerful tool whereby a normal gene can be introduced into patient's cells to correct an inherited defect. In patients with a hematologic genetic defect, the type of defect determines which target cell, vector and gene are used. In theory, all of the genetic defects treated by hematopoietic stem cell transplantation (HSCT) would benefit from gene transfer into HSC, while genetic defects treated by protein replacement would benefit from in vivo gene transfer into a cell able to export the synthetized protein. In practice, this distinction is only partially true because β-thalassemic symptoms are improved by continuous erythropoietin delivery from muscle,[102] and adenosine deaminase (ADA) deficiency may benefit from gene transfer into HSC.[103]

As HSC are easily harvested and manipulated ex vivo, they seem to be ideal targets for gene transfer. However, in practice, this assumption must be modulated: (1) stem cells have not been identified in human beings but only in mice,[104] (2) HSC cannot be expanded without loss of engraftment potential,[105] (3) human HSC is poorly transduced by oncoretrovirus vectors, which only integrate if the target cells proliferate extensively,[106] and, (4) as some genetic diseases of the lymphohematopoietic system involve tight regulation or lineage specific expression of the mutated protein or both, further improvements in gene therapy tools are needed to treat these diseases using this approach.[107]

Nevertheless, progress in basic hematopoiesis, immunological mechanisms, vector design, and molecular insight into disease should lead to improvements in gene therapy for patients with genetic blood diseases in the near future. A recent clinical trial in patients affected by severe combined immunodeficiency disease provides the proof of principle that this new approach can be beneficial for the patients.

Severe Combined Immunodeficiency Disease

Severe combined immunodeficiency disease (SCID) is a group of diseases characterized by impaired T lymphocyte development. Several conditions have been described according to the cell and gene affected, and the modality of inheritance. To date, nine SCID genes have been identified;[16,108-111] however, the molecular mechanism of reticular dysgenesis and some T (-) B (+) NK (+) SCID remains unknown.

SCID is an ideal candidate for gene therapy because it is the most severe form of primary immunodeficiencies (PID), it is lethal within the first year of life, monogenic, and characterized by an early block in T cell pathway differentiation. Relieving the block should provide a growth advantage to the T progenitor cells. Furthermore, toxic effects should not occur due to the nonregulated expression of the transgene, as the gene involved may be a housekeeping gene, such as in adenosine deaminase deficiency (ADA), or a protein regulated by otherwise normal subunits such as in the γc deficiency. Lastly, allogeneic stem cell transplantation in the absence of a HLA genoidentical donor is only partially successful. Although a recent study showed a 78% survival rate in patients without GVHD, a severe deficiency in B cells is common in these long-term survivors, and involves long-term treatment with immunoglobulins.[16,112,113]

Adenosine Deaminase Deficiency

Adenosine deaminase (ADA) deficiency was the first disease to be treated by gene therapy as this gene was among the first genes identified. The adenosine deaminase enzyme is responsible for detoxification of metabolites in the purine salvage pathway. Six clinical trials have been conducted since 1990. The first protocol consisted of repeated infusions of peripheral T cells transduced ex vivo with an ADA-cDNA retroviral vector.[114] Of the other three clinical trials which targeted autologous HSC,[115-117] only the study of Kohn et al showed unequivocally the selective advantage conferred in vivo by ADA transgene expression in T lymphocytes. In fact, transgene-containing cells ranged from 1 to 10% in the T lymphocytes versus only 0.01% to 0.1% in the other hematopoietic lineages. Despite these encouraging results, a functional immune system was not restored due to insufficient expression of the ADA-gene. Infants and children with ADA-deficient SCID have transgene containing peripheral blood lymphocytes more than seven years after treatment without any adverse effects.[108,114] Three major changes in the gene therapy protocol have significantly improved these clinical results leading to restoration of the clinical phenotype. Stop of the PEG-ADA substitutive treatment administered simultaneously to the modified hematopoietic precursor cells, combined with the use of a mild conditioning regimen and of an optimized ex vivo transduction protocol based on the use a human recombinant fragment of fibronectin and early-acting hematopoietic cytokines such as Flt3-L and MGDF have allowed a complete restoration of immunological compartment in one out of two treated patients.[103] The role played by the conditioning regimen in this achievement seems to be of primary importance.

X-Linked SCID

The X-linked SCID (SCID-XI) is caused by mutations in the γc encoding gene and it accounts for 50-60 % of patients with SCID. The γc chain is shared by several hematopoietic cytokine receptors including the IL-2, IL-4, IL-7, IL-9, IL-15 and IL-21 receptors (R),[118,119] which explains why T/NK lymphoid lineages are absent in this disease. Several studies in mutant mice have thrown light on the key role of IL-7 / IL-7R binding in inducing survival and proliferation of early T cell progenitors in the thymus.[120,121] These findings were confirmed by the absence of T cell development in two patients with IL-7R α deficiencies.[122,123] The NK cell deficiency is due to a defect in IL-15 induced signaling,[124,125] even though IL-21 may play a role in the function of these cells. As γc belongs to the hematopoietic cytokine receptor family, and is constitutively expressed by T, B and NK cells, as well as myeloid cells and erythroblasts,[126] it is likely that nonregulated expression is not toxic for hematopoiesis. γc gene transfer studies in animals are consistent with this hypothesis.[127-129]

On the basis of extensive preclinical data accumulated by our team[125,130,131] and other groups[129,132,133] supporting the use of gene therapy to treat patients with SCID-XI, the first five patients were enrolled in a clinical trial in March 1999, and approval for a further six patients was recently obtained. This trial was proposed for SCID-XI patients lacking an HLA-genoidentical donor. The protocol consisted of marrow harvesting, CD34-cell purification, cell preactivation in the presence of SCF, FLT3-L, MGDF and IL-3, followed by 3 cycles of infection with the supernatant containing the defective retroviral vector in a bag coated with the CH-296 fibronectin fragment over 3 days.[134] Clinical results of the first five treated patients were reported recently.[135,136] A more than two-year follow-up of these five patients has confirmed the preliminary data, which showed a favorable clinical outcome. Transduced T cells, and, to some extent, NK cells developed within 3-4 months, which led to the development of both T and B cell antigen-specific responses. So far nine patients have been treated in our hospital and in light of them the infusion of transduced CD34+ cells resulted in the generation of peripheral transduced T-cells with characteristics similar to those of age-matched controls in terms of cell count, subset distribution, TCR diversity and antigen-driven activation within 6-12 weeks. Correction of T-cell immunodeficiency has been sustained for a period of up to 3 years. Furthermore, despite a low rate of B-cell transduction, immunoglobulin production was at least in part restored. If the correction of humoral response is sustained, this treatment could be proposed to patients who have been transplanted with a haploidentical related donor and who lack B-cell reconstitution. It may also provide insight into the absence of T and B cell cooperation in haploidentical HSCT setting. To date, the first treated children have normal serum levels of IgG, IgA and IgM and have not required immunoglobulins substitution more than two years after the treatment.[136]

However, although these findings provide evidence that gene therapy can correct the immunological phenotype of SCID-X1, a much longer period of observation is needed to determine, in one hand, how long this effect persists and, in the other, its safety. Several possible mechanisms including silencing of the transgene as in ADA-deficient patients,[115] or a decline in transduced precursors cells and thymic function,[137] could lead to a decrease in clinical benefit over time. Sequential analysis of the provirus integration sites in lymphocyte and myeloid cells using the method described by Von Kalle et al are currently underway to answer two key questions: how immature is the hematopoietic cell transduced by this protocol and how great is the risk of insertional mutagenesis.[138] This last concern is raised from the recent observation of a gamma/delta T cell monoclonal lymphoproliferation developed in one child treated three years ago. Molecular studies revealed that all the γδ T-cells have a single provirus integration site in the intron of the LMO-2 gene, which is aberrantly expressed in these cells. This theoretical risk was aware but it seemed very small, as this phenomenon did not occur in animal experiments.[127,128] Its understanding should shed light if this is a very unluckily random event or if this risk has been under-estimated in the past. In the meanwhile, this protocol is hold up at least temporarily.

Other SCID conditions, including Jak-3, IL-7Rα, and RAG-1 and RAG-2 deficiencies, could benefit from the same selective growth advantage observed in the SCID-X1 clinical trial and their development is partially dependant on a deep appreciation of risk/benefit ratio in the gene therapy approach.[108,123,139] The specific association of the γc receptor subunit with the tyrosine kinase Jak 3 explains how mutations in the Jak-3 encoding gene results in a SCID condition with a strictly identical clinical phenotype to that of the γc gene mutation.[108] In vitro gene transfer studies using lymphocytes from Jak-3 deficient SCID patients have shown that it is possible to restore normal cytokine signaling.[140] A recent report on the correction of immune function in Jak 3 knock-out mice without any conditioning regimen provides further evidence for the use of gene therapy in this SCID disorder.[141]

The great concern with this disease is the potential toxic risk due the uncontrolled expression of a tyrosine kinase in the hematopoietic system and to the positional effect. To date, mouse models have not shown any undesirable effects. Similarly, patients lacking expression of either RAG1 or RAG2 proteins suffer from a SCID condition characterized by a lack of T and B lymphocytes. RAG1 and RAG2 genes are essential components of the V(D) J recombination leading to the formation of B and T cell receptor diversity. We are currently considering ex vivo gene therapy of hematopoietic stem cells as an alternative to partially incompatible HSCT, as a selective advantage of transgene expressing cells on endogenous RAG deficient cells is expected. Moreover, constitutive expression of only one of the two RAG proteins should not be harmful since concomitant expression of both genes is required for recombination activity. Murine RAG1 and RAG2 deficient models, which exhibit the same phenotype as RAG$^{-/-}$ patients, will allow us to test these hypotheses. Correction of RAG1 or RAG2 deficiencies by gene transfer in mice will constitute a step towards clinical application.[142]

Fanconi Anemia

Fanconi anemia (FA) is an autosomal recessive disorder characterized clinically by bone marrow failure, multiple congenital abnormalities and susceptibility to cancer.[143] Complementation analysis of FA cells using somatic cell fusion has identified at least eight different forms. The common feature of this condition is the hypersensitivity of FA cells to DNA damage by interstrand cross-linking agents, which varies greatly between complementation groups and families. Regardless of the group, most patients with FA die from complications due to bone marrow failure. Identification of the FA genes may provide an alternative therapeutic approach in patients without a histocompatible donor.[144] Given that the 3-year survival rate using HLA-matched unrelated donors is only 33%,[145] the search for alternative treatments is warranted. Recent reports of in vivo selection of wild-type hematopoietic stem cells in a murine model of FA,[146] and in one of two affected siblings,[147] provide further support for the use of gene therapy to obtain a selective advantage of transduced cells. A recent study showed that the presence of 74-87% of reverted cells was sufficient to maintain adequate, although not normal, hematopoiesis at age 16, whereas the affected sibling had severe bone marrow failure by 6.8 years of age.[147] This observation raises a number of questions because, despite the partially-reverted phenotype, the child developed a clonal abnormality within the nonreverted FA population, which indicates that complete eradication of the FA cell population is necessary in this setting. Nevertheless, despite the development of this clonal abnormality, the partial correction of hematopoiesis indicates the benefit of gene therapy.

A clinical trial has been initiated at NIH and four patients have been enrolled.[148] Three of the four patients have been successfully mobilized, despite the low number of HSC present in the bone marrow of these patients. PCR+ CFU-GM colonies which were resistant to treatment with mitomycin-C (MMC) were detected in two of the patients several months after the gene therapy. Of note, transduced positive cells were detected in one patient only after irradiation for a gynecologic malignancy. A nonoptimized transduction protocol of HSC associated with late harvest of HSC and the absence of in vivo selection by a pharmacological agent or a cytokine to amplify the survival advantage of the transduced cells may explain the absence of clinical benefit. The latter may be essential to ensure the success of gene therapy in FA as the same group recently reported the hematopoietic correction of FANCA deficient mice by in vivo administration of MMC.[149] Finally, Rio et al recently reported for the first time that the transduction of murine FANCA deficient progenitors with vectors encoding for the human FANCA gene reverts at least in vitro their hypersensitivity to MMC.[150,151] These experiments not only demonstrate the efficacy of retroviral vectors for correcting a characteristic FA phenotype but also demonstrate the applicability of FANCA deficient mice for assessing the efficacy of vectors encoding the human homologous gene.

Chronic Granulomatous Disease

Chronic granulomatous disease (CGD) is an inherited immune deficiency characterized by failure of respiratory burst and impaired anti-microbial activity due to defects in any one of the four subunits of the phagocyte NADPH oxidase (phox).[150] The incidence is approximately 1 in 250,000 individuals, and the most common form (about two thirds of cases) is X-chromosome linked resulting from mutations in the gp 91 phox gene. CGD patients are predisposed to recurrent and often life-threatening bacterial and fungal infections.[150,152] Despite adequate antibiotic and antifungal lifelong prophylaxis, patients are still susceptible to life-threatening infections especially from pathogens such as Aspergillus spp. A recent study in 368 patients estimates the mortality rate at approximately 5% per year for patients with the X-linked recessive form of the disease, and 2% per year for those with the autosomal recessive form. When an HLA-identical donor is available, HSCT is an alternative approach in selected patients. Nevertheless, the choice of "when" HSCT becomes a good therapeutical option for patients with recurrent serious infection despite adequate prophylaxis is still debated due to transplantation-related toxicity, especially when HSCT is performed in patients with invasive infection, multiple liver abscesses, and inflammatory sequelae.[153,154]

Thus, for "high risk" patients without an HLA-identical donor (i.e., 70% of cases), gene therapy should provide a good alternative. However, in contrast to SCID patients, one obstacle to this approach is that corrected CGD cells have no selective growth advantage in vivo over the defective, nontransduced counterparts. Nevertheless, as little as 5-8% of oxidase positive cells may be sufficient to improve the clinical status of these patients, given the findings in asymptotic X-CGD carriers,[155] and in animal models.[156,157] Moreover, in patients without stable neutrophile correction, gene therapy may be beneficial as a short-term adjuvant therapy for life-threatening infections. The feasibility of genetic correction of different forms of CGD has been demonstrated in vivo after transplantation of genetically corrected bone marrow cells into gp 91^{phox} or p47^{phox} deficient CGD mice.[157,158]

In contrast to the promising results in mice, the results of the NIH clinical phase I trial were disappointing.[159] In this protocol, autologous peripheral blood CD34$^+$ cells were transduced with p47^{phox} cDNA containing retroviral vector and then infused. Peripheral blood neutrophiles with respiratory burst oxidase activity were seen for up to 3 to 6 months in all five patients studied, although the frequency of oxidase-positive neutrophiles was 0.02% to 0.005%. The most likely explanation for the discrepancy between the human and mouse studies was that the patients did not receive pretransplant conditioning, whereas the mice were subjected to high-dose irradiation. A "mild" conditioning regimen and improved CD34$^+$ transduction should lead to the achievement the therapeutic level of stable 5% transduced neutrophils in the near future.

Currently, two approaches to increase the transduction efficiency of CD34$^+$ cells are being investigated. First, the transduction efficiency of the MFG-based retrovirus can be improved by changing the envelope. Recent studies have shown that human stem cells express a low level of receptor for the amphotropic envelope, whereas they express high levels of the receptor for RD 114, the feline endogenous retrovirus envelope[160] or Gibbon ape leukemia virus. Second HIV-based vectors have been shown to circumvent the requirement of cell cycling to integrate oncoretrovirus vectors.[161] A recent study of in vitro transduction of human X-CGD cells showed a correction of CGD phenotype after gene transfer by a lentiviral vector.[162,163] Finally, preclinical and clinical studies have demonstrated that in vivo selection of transduced cells increases corrected neutrophiles to the required therapeutic level.[164,165] Clinical trials in patients with CGD are currently underway in Europe and the United States; and these should provide new insights into ways to improve the therapeutic approach to this inherited disease.

Hemoglobinopathies

Thalassemia and sickle cell disease were the earliest monogenic disorders considered for gene therapy. These hemoglobinopathies are associated with high morbidity and mortality. Severe β-thalassemia is characterized by ineffective erythropoiesis and hemolytic anemia, which need lifelong red cell transfusions. Despite recent improvement in patient survival, β-thalassemia remains a serious public health problem.

To date, β-globin gene transfer into HSC in mice has been too low to warrant investigation in a clinical trial. The most important technical challenge has been to design vectors containing the functional globin gene and the regulatory elements necessary to achieve high-level and stable globin gene expression in developing erythroblasts.

Initial reports have indicated that despite the inclusion of elements from the locus control region (LCR) into the retroviral vectors, the level of β-globin expression decreased over time, suggesting complete silencing of the vector-encoded globin gene.[166,167] Another major obstacle to achieve sustained expression of transduced globin genes in differentiating erythroblasts is the risk of position effect variegation (PEV) of expression, which is a reflection of the effects of flanking chromatin in differentiated cells and of chromatin remodeling at the site of integration in the progeny of pluripotential cells. To avoid in vivo stem cell β-globin gene silencing, ex vivo preselection of retrovirally transduced stem cells using GFP expression (green fluorescent protein gene marker) juxtaposed to the β-globin/LCR segment was investigated.[168,169] All mice engrafted with preselected cells showed sustained expression of human β-globin and GFP in red cells.

Lentiviral vectors were recently successfully exploited to achieve efficient and stable transfer of large β-globin gene/LCR segments resulting in therapeutical levels of β-globin gene expression in both β-thalassemia[170-173] and sickle cell disease (SCD) mouse models.[174] In particular, the data published by Pawliuk[174] based on the use of an anti-sickling lentiviral vector show that virtually all HSC from the graft appeared transduced with several integrated proviruses per cell conversely to the data published by May.[172] These differences are likely due to the different lentiviral vector as further proved in the Imren's work.[173]

Conclusions and Perspectives in the Use of HSC for Regenerative Medicine

To complete this overview on the therapeutical capacities of bone marrow derived stem cells, some recently published data have to be mentioned. These data are based on the observation that stem cells, particularly those from bone marrow, have the capacity of colonizing different tissues and transdifferentiate into cell lineages of the organ.[175-179] In particular it has been reported in experimental settings that bone marrow derived stem cells can repair an infarcted heart as well as a genetic liver disease.[116,180,181]

These data have to be interpreted with caution and they require further experimental work before claiming that adult derived stem cells can be used to cure a number of genetic or acquired diseases. Nevertheless they have changed old "dogmas" on boundaries and differentiation capacity of adult stem cells.

References

1. Sullivan KM, Parkman R, Walters MC. Bone Marrow Transplantation for NonMalignant Disease. Hematology (Am Soc Hematol Educ Program) 2000; 319-38.
2. Gratwohl A, Baldomero H, Horisberger B et al. Current trends in hematopoietic stem cell transplantation in Europe. Blood 2002; 100(7):2374-86.
3. Anasetti C, Amos D, Beatty PG et al. Effect of HLA compatibility on engraftment of bone marrow transplants in patients with leukemia or lymphoma. N Engl J Med 1989; 320(4):197-204.
4. Petersdorf EW, Gooley TA, Anasetti C et al. Optimizing outcome after unrelated marrow transplantation by comprehensive matching of HLA class I and II alleles in the donor and recipient. Blood 1998; 92(10):3515-20.
5. Petersdorf EW, Longton GM, Anasetti C et al. Association of HLA-C disparity with graft failure after marrow transplantation from unrelated donors. Blood 1997; 89(5):1818-23.
6. Petersdorf EW, Kollman C, Hurley CK et al. Effect of HLA class II gene disparity on clinical outcome in unrelated donor hematopoietic cell transplantation for chronic myeloid leukemia: the US National Marrow Donor Program Experience. Blood 2001; 98(10):2922-9.
7. Sasazuki T, Juji T, Morishima Y et al. Effect of matching of class I HLA alleles on clinical outcome after transplantation of hematopoietic stem cells from an unrelated donor. Japan Marrow Donor Program N Engl J Med 1998; 339(17):1177-85.
8. Drobyski WR, Klein J, Flomenberg N et al. Superior survival associated with transplantation of matched unrelated versus one-antigen-mismatched unrelated or highly human leukocyte antigen-disparate haploidentical family donor marrow grafts for the treatment of hematologic malignancies: establishing a treatment algorithm for recipients of alternative donor grafts. Blood 2002; 99(3):806-14.
9. Caillat-Zucman S, Haddad E, Fischer A et al. Similar outcome after transplantation of bone marrow from genoidentical and perfectly matched unrelated donors: a pediatric single-center study. Blood 1999; 96(11):476a.
10. Antoine C, Muller S, Cant A et al. Long term survival and hematopoietic stem cell transplantation for immunodeficiencies: Report of the european experience 1968-1999. Lancet 2003; 361(9357):553-560.

11. Fischer A, Landais P, Friedrich W et al. European experience of bone-marrow transplantation for severe combined immunodeficiency. Lancet 1990; 336(8719):850-4.
12. Fischer A, Landais P, Friedrich W et al. Bone marrow transplantation (BMT) in Europe for primary immunodeficiencies other than severe combined immunodeficiency: a report from the European Group for BMT and the European Group for Immunodeficiency. Blood 1994; 83(4):1149-54.
13. Henslee-Downey PJ, Abhyankar SH, Parrish RS et al. Use of partially mismatched related donors extends access to allogeneic marrow transplant. Blood 1997; 89(10):3864-72.
14. Haddad E, Landais P, Friedrich W et al. Long-term immune reconstitution and outcome after HLA-nonidentical T- cell-depleted bone marrow transplantation for severe combined immunodeficiency: a European retrospective study of 116 patients. Blood 1998; 91(10):3646-53.
15. Aversa F, Tabilio A, Velardi A et al. Treatment of high-risk acute leukemia with T-cell-depleted stem cells from related donors with one fully mismatched HLA haplotype. N Engl J Med 1998; 339(17):1186-93.
16. Buckley RH, Schiff SE, Schiff RI et al. Hematopoietic stem-cell transplantation for the treatment of severe combined immunodeficiency. N Engl J Med 1999; 340(7):508-16.
17. Storek J, Gooley T, Witherspoon RP et al. Infectious morbidity in long-term survivors of allogeneic marrow transplantation is associated with low CD4 T cell counts. Am J Hematol 1997; 54(2):131-8.
18. Small TN, Papadopoulos EB, Boulad F et al. Comparison of immune reconstitution after unrelated and related T-cell-depleted bone marrow transplantation: effect of patient age and donor leukocyte infusions. Blood 1999; 93(2):467-80.
19. Bensinger WI, Clift R, Martin P et al. Allogeneic peripheral blood stem cell transplantation in patients with advanced hematologic malignancies: a retrospective comparison with marrow transplantation. Blood 1996; 88(7):2794-800.
20. Schmitz N, Bacigalupo A, Labopin M et al. Transplantation of peripheral blood progenitor cells from HLA-identical sibling donors. European Group for Blood and Marrow Transplantation (EBMT) Br J Haematol 1996; 95(4):715-23.
21. Powles R, Mehta J, Kulkarni S et al. Allogeneic blood and bone-marrow stem-cell transplantation in haematological malignant diseases: a randomised trial. Lancet 2000; 355(9211):1231-7.
22. Bensinger WI, Martin PJ, Storer B et al. Transplantation of bone marrow as compared with peripheral-blood cells from HLA-identical relatives in patients with hematologic cancers. N Engl J Med 2001; 344(3):175-81.
23. Storek J, Gooley T, Siadak M et al. Allogeneic peripheral blood stem cell transplantation may be associated with a high risk of chronic graft-versus-host disease. Blood 1997; 90(12):4705-9.
24. Scott MA, Gandhi MK, Jestice HK et al. A trend towards an increased incidence of chronic graft-versus-host disease following allogeneic peripheral blood progenitor cell transplantation: a case controlled study. Bone Marrow Transplant 1998; 22(3):273-6.
25. Blaise D, Kuentz M, Fortanier C et al. Randomized trial of bone marrow versus lenograstim-primed blood cell allogeneic transplantation in patients with early-stage leukemia: a report from the Societe Francaise de Greffe de Moelle. J Clin Oncol 2000; 18(3):537-46.
26. Morton J, Hutchins C, Durrant S. Granulocyte-colony-stimulating factor (G-CSF)-primed allogeneic bone marrow: significantly less graft-versus-host disease and comparable engraftment to G-CSF-mobilized peripheral blood stem cells. Blood 2001; 98(12):3186-91.
27. Champlin RE, Schmitz N, Horowitz MM et al. Blood stem cells compared with bone marrow as a source of hematopoietic cells for allogeneic transplantation. IBMTR Histocompatibility and Stem Cell Sources Working Committee and the European Group for Blood and Marrow Transplantation (EBMT). Blood 2000; 95(12):3702-9.
28. Couban S, Messner HA, Andreou P et al. Bone marrow mobilized with granulocyte colony-stimulating factor in related allogeneic transplant recipients: a study of 29 patients. Biol Blood Marrow Transplant 2000; 6(4A):422-7.
29. Isola L, Scigliano E, Fruchtman S. Long-term follow-up after allogeneic granulocyte colony-stimulating factor—primed bone marrow transplantation. Biol Blood Marrow Transplant 2000; 6(4A):428-33.
30. Serody JS, Sparks SD, Lin Y et al. Comparison of granulocyte colony-stimulating factor (G-CSF)— mobilized peripheral blood progenitor cells and G-CSF—stimulated bone marrow as a source of stem cells in HLA-matched sibling transplantation. Biol Blood Marrow Transplant 2000; 6(4A):434-40.

31. Ringden O, Remberger M, Runde V et al. Faster engraftment of neutrophils and platelets with peripheral blood stem cells from unrelated donors: a comparison with marrow transplantation. Bone Marrow Transplant 2000; 25(Suppl 2):S6-8.

32. Remberger M, Ringden O, Blau IW et al. No difference in graft-versus-host disease, relapse, and survival comparing peripheral stem cells to bone marrow using unrelated donors. Blood 2001; 98(6):1739-45.

33. Fauser AA, Basara N, Blau IW et al. A comparative study of peripheral blood stem cell vs bone marrow transplantation from unrelated donors (MUD): a single center study. Bone Marrow Transplant 2000; 25(Suppl 2):27-31.

34. Elmaagacli AH, Beelen DW, Opalka B et al. The risk of residual molecular and cytogenetic disease in patients with Philadelphia-chromosome positive first chronic phase chronic myelogenous leukemia is reduced after transplantation of allogeneic peripheral blood stem cells compared with bone marrow. Blood 1999; 94(2):384-9.

35. Champlin RE, Passweg JR, Zhang MJ et al. T-cell depletion of bone marrow transplants for leukemia from donors other than HLA-identical siblings: advantage of T-cell antibodies with narrow specificities. Blood 2000; 95(12):3996-4003.

36. Conneally E, Cashman J, Petzer A et al. Expansion in vitro of transplantable human cord blood stem cells demonstrated using a quantitative assay of their lympho-myeloid repopulating activity in nonobese diabetic-scid/scid mice. Proc Natl Acad Sci USA 1997; 94(18):9836-41.

37. Piacibello W, Sanavio F, Severino A et al. Engraftment in nonobese diabetic severe combined immunodeficient mice of human CD34(+) cord blood cells after ex vivo expansion: evidence for the amplification and self-renewal of repopulating stem cells. Blood 1999; 93(11):3736-49.

38. Denning-Kendall PA, Evely R, Singha S et al. In vitro expansion of cord blood does not prevent engraftment of severe combined immunodeficient repopulating cells. Br J Haematol 2002; 116(1):218-28.

39. Wagner JE, Kernan NA, Steinbuch M et al. Allogeneic sibling umbilical-cord-blood transplantation in children with malignant and nonmalignant disease. Lancet 1995; 346(8969):214-9.

40. Kurtzberg J, Laughlin M, Graham ML et al. Placental blood as a source of hematopoietic stem cells for transplantation into unrelated recipients. N Engl J Med 1996; 335(3):157-66.

41. Wagner JE, Rosenthal J, Sweetman R et al. Successful transplantation of HLA-matched and HLA-mismatched umbilical cord blood from unrelated donors: analysis of engraftment and acute graft-versus-host disease. Blood 1996; 88(3):795-802.

42. Gluckman E, Rocha V, Boyer-Chammard A et al. Outcome of cord-blood transplantation from related and unrelated donors. Eurocord Transplant Group and the European Blood and Marrow Transplantation Group. N Engl J Med 1997; 337(6):373-81.

43. Rocha V, Wagner Jr JE, Sobocinski KA et al. Graft-versus-host disease in children who have received a cord-blood or bone marrow transplant from an HLA-identical sibling. Eurocord and International Bone Marrow Transplant Registry Working Committee on Alternative Donor and Stem Cell Sources. N Engl J Med 2000; 342(25):1846-54.

44. Barker JN, Davies SM, DeFor T et al. Survival after transplantation of unrelated donor umbilical cord blood is comparable to that of human leukocyte antigen-matched unrelated donor bone marrow: results of a matched-pair analysis. Blood 2001; 97(10):2957-61.

45. Rocha V, Cornish J, Sievers EL et al. Comparison of outcomes of unrelated bone marrow and umbilical cord blood transplants in children with acute leukemia. Blood 2001; 97(10):2962-71.

46. Rubinstein P, Carrier C, Scaradavou A et al. Outcomes among 562 recipients of placental-blood transplants from unrelated donors. N Engl J Med 1998; 339(22):1565-77.

47. Laughlin MJ, Barker J, Bambach B et al. Hematopoietic engraftment and survival in adult recipients of umbilical-cord blood from unrelated donors. N Engl J Med 2001; 344(24):1815-22.

48. Ohishi K, Varnum-Finney B, Bernstein ID. Delta-1 enhances marrow and thymus repopulating ability of human CD34(+)CD38(-) cord blood cells. J Clin Invest 2002; 110(8):1165-74.

49. Dicke KA, Tridente G, van Bekkum DW. The selective elimination of immunologically competent cells from bone marrow and lymphocyte cell mixtures. 3 In vitro test for detection of immunocompetent cells in fractionated mouse spleen cell suspensions and primate bone marrow suspensions. Transplantation 1969; 8(4):422-34.

50. Reisner Y, Itzicovitch L, Meshorer A et al. Hemopoietic stem cell transplantation using mouse bone marrow and spleen cells fractionated by lectins. Proc Natl Acad Sci USA 1978; 75(6):2933-6.

51. Korngold R, Sprent J. Negative selection of T cells causing lethal graft-versus-host disease across minor histocompatibility barriers. Role of the H-2 complex. J Exp Med 1980; 151(5):1114-24.

52. Ho VT, Soiffer RJ. The history and future of T-cell depletion as graft-versus-host disease prophylaxis for allogeneic hematopoietic stem cell transplantation. Blood 2001; 98(12):3192-204.

53. Waldmann H, Polliak A, Hale G et al. Elimination of graft-versus-host disease by in vitro depletion of alloreactive lymphocytes with a monoclonal rat anti-human lymphocyte antibody (CAMPATH-1). Lancet 1984; 2(8401):483-6.

54. Reisner Y, Kapoor N, Kirkpatrick D et al. Transplantation for acute leukaemia with HLA-A and B nonidentical parentalarrow cells fractionated with soybean agglutinin and sheep red blood cells. Lancet 1981; 2(8242):327-31.

55. Dreger P, Viehmann K, Steinmann J et al. G-CSF-mobilized peripheral blood progenitor cells for allogeneic transplantation: comparison of T cell depletion strategies using different CD34+ selection systems or CAMPATH-1. Exp Hematol 1995; 23(2):147-54.

56. Watts MJ, Somervaille TC, Ings SJ et al. Variable product purity and functional capacity after CD34 selection: a direct comparison of the CliniMACS (v2.1) and Isolex 300i (v2.5) clinical scale devices. Br J Haematol 2002; 118(1):117-23.

57. Zaucha JM, Gooley T, Bensinger WI et al. CD34 cell dose in granulocyte colony-stimulating factor-mobilized peripheral blood mononuclear cell grafts affects engraftment kinetics and development of extensive chronic graft-versus-host disease after human leukocyte antigen-identical sibling transplantation. Blood 2001; 98(12):3221-7.

58. Uchida N, Tsukamoto A, He D et al. High doses of purified stem cells cause early hematopoietic recovery in syngeneic and allogeneic hosts. J Clin Invest 1998; 101(5):961-6.

59. Reisner Y, Bachar-Lustig E, Li HW. Purified Sca1+Lin- stem cells can tolerize fully allogeneic host T-cells remaining after sublethal TBI. 1997.

60. Goodell MA. Introduction: Focus on hematology. CD34(+) or CD34(-): does it really matter? Blood 1999; 94(8):2545-7.

61. Bhatia M, Bonnet D, Murdoch B et al. A newly discovered class of human hematopoietic cells with SCID-repopulating activity. Nat Med 1998; 4(9):1038-45.

62. Bianco P, Gehron Robey P. Marrow stromal stem cells. J Clin Invest 2000; 105(12):1663-8.

63. Cavazzana-Calvo M, Bensoussan D, Jabado N et al. Prevention of EBV-induced B-lymphoproliferative disorder by ex vivo marrow B-cell depletion in HLA-phenoidentical or nonidentical T-depleted bone marrow transplantation. Br J Haematol 1998; 103(2):543-51.

64. Steidl U, Kronenwett R, Rohr UP et al. Gene expression profiling identifies significant differences between the molecular phenotypes of bone marrow-derived and circulating human CD34+ hematopoietic stem cells. Blood 2002; 99(6):2037-44.

65. Gabbianelli M, Pelosi E, Montesoro E et al. Multi-level effects of flt3 ligand on human hematopoiesis: expansion of putative stem cells and proliferation of granulomonocytic progenitors/monocytic precursors. Blood 1995; 86(5):1661-70.

66. Glimm H, Eaves CJ. Direct evidence for multiple self-renewal divisions of human in vivo repopulating hematopoietic cells in short-term culture. Blood 1999; 94(7):2161-8.

67. Gupta P, Oegema TR Jr. et al. Human LTC-IC can be maintained for at least 5 weeks in vitro when interleukin-3 and a single chemokine are combined with O-sulfated heparan sulfates: requirement for optimal binding interactions of heparan sulfate with early-acting cytokines and matrix proteins. Blood 2000; 95(1):147-55.

68. Kobayashi M, Laver JH, Kato T et al. Thrombopoietin supports proliferation of human primitive hematopoietic cells in synergy with steel factor and/or interleukin-3. Blood 1996; 88(2):429-36.

69. Briddell RA, Hartley CA, Smith KA et al. Recombinant rat stem cell factor synergizes with recombinant human granulocyte colony-stimulating factor in vivo in mice to mobilize peripheral blood progenitor cells that have enhanced repopulating potential. Blood 1993; 82(6):1720-3.

70. Elias AD, Ayash L, Anderson KC et al. Mobilization of peripheral blood progenitor cells by chemotherapy and granulocyte-macrophage colony-stimulating factor for hematologic support after high-dose intensification for breast cancer. Blood 1992; 79(11):3036-44.

71. Sudo Y, Shimazaki C, Ashihara E et al. Synergistic effect of FLT-3 ligand on the granulocyte colony-stimulating factor-induced mobilization of hematopoietic stem cells and progenitor cells into blood in mice. Blood 1997; 89(9):3186-91.

72. Wendling F, Maraskovsky E, Debili N et al. cMpl ligand is a humoral regulator of megakaryocytopoiesis. Nature 1994; 369(6481):571-4.

73. McNiece I, Jones R, Bearman SI et al. Ex vivo expanded peripheral blood progenitor cells provide rapid neutrophil recovery after high-dose chemotherapy in patients with breast cancer. Blood 2000; 96(9):3001-7.

74. Reiffers J, Cailliot C, Dazey B et al. Abrogation of post-myeloablative chemotherapy neutropenia by ex-vivo expanded autologous CD34-positive cells. Lancet 1999; 354(9184):1092-3.

75. Norol F, Drouet M, Mathieu J et al. Ex vivo expanded mobilized peripheral blood CD34+ cells accelerate haematological recovery in a baboon model of autologous transplantation. Br J Haematol 2000; 109(1):162-72.

76. Antonchuk J, Sauvageau G, Humphries RK. HOXB4 overexpression mediates very rapid stem cell regeneration and competitive hematopoietic repopulation. Exp Hematol 2001; 29(9):1125-34.

77. Sauvageau G, Thorsteinsdottir U, Eaves CJ et al. Overexpression of HOXB4 in hematopoietic cells causes the selective expansion of more primitive populations in vitro and in vivo. Genes Dev 1995; 9(14):1753-65.

78. Antonchuk J, Sauvageau G, Humphries RK. HOXB4-induced expansion of adult hematopoietic stem cells ex vivo. Cell 2002; 109(1):39-45.

79. Daga A, Podesta M, Capra MC et al. The retroviral transduction of HOXC4 into human CD34(+) cells induces an in vitro expansion of clonogenic and early progenitors. Exp Hematol 2000; 28(5):569-74.

80. McGuire WP. High-dose chemotherapy and autologous bone marrow or stem cell reconstitution for solid tumors. Curr Probl Cancer 1998; 22(3):135-77.

81. Rahman ZU, Frye DK, Buzdar AU et al. Impact of selection process on response rate and long-term survival of potential high-dose chemotherapy candidates treated with standard-dose doxorubicin-containing chemotherapy in patients with metastatic breast cancer. J Clin Oncol 1997; 15(10):3171-7.

82. Garcia-Carbonero R, Hidalgo M, Paz-Ares L et al. Patient selection in high-dose chemotherapy trials: relevance in high-risk breast cancer. J Clin Oncol 1997; 15(10):3178-84.

83. van der Wall E, Horn T, Bright E et al. Autologous graft-versus-host disease induction in advanced breast cancer: role of peripheral blood progenitor cells. Br J Cancer 2000; 83(11):1405-11.

84. Ahmed T, Kancherla R, Qureshi Z et al. High-dose chemotherapy and stem cell transplantation for patients with stage IV breast cancer without clinically evident disease: correlation of CD34+ selection to clinical outcome. Bone Marrow Transplant 2000; 25(10):1041-5.

85. Stadtmauer EA, O'Neill A, Goldstein LJ et al. Conventional-dose chemotherapy compared with high-dose chemotherapy plus autologous hematopoietic stem-cell transplantation for metastatic breast cancer. Philadelphia Bone Marrow Transplant Group. N Engl J Med 2000; 342(15):1069-76.

86. Lippman ME. High-dose chemotherapy plus autologous bone marrow transplantation for metastatic breast cancer. N Engl J Med 2000; 342(15):1119-20.

87. Joske DJ, Ma DT, Langlands DR et al. Autologous bone-marrow transplantation for rheumatoid arthritis. Lancet 1997; 350(9074):337-8.

88. Tyndall A, Black C, Finke J et al. Treatment of systemic sclerosis with autologous haemopoietic stem cell transplantation. Lancet 1997; 349(9047):254.

89. Tyndall A, Gratwohl A. Bone marrow transplantation in the treatment of autoimmune diseases. Br J Rheumatol 1997; 36(1):1-3.

90. Marmont AM, van Lint MT, Gualandi F et al. Autologous marrow stem cell transplantation for severe systemic lupus erythematosus of long duration. Lupus 1997; 6(6):545-8.

91. Brooks PM, Atkinson KA, Hamilton JA. Stem cell transplantation in autoimmune disease. J Rheumatol 1995; 22:1809-11.

92. Fassas A, Anagnostopoulos A, Kazis A et al. Peripheral blood stem cell transplantation in the treatment of progressive multiple sclerosis: first results of a pilot study. Bone Marrow Transplant 1997; 20(8):631-8.

93. Wulffraat N, van Royen A, Bierings M et al. Autologous haemopoietic stem-cell transplantation in four patients with refractory juvenile chronic arthritis. Lancet 1999; 353(9152):550-3.

94. Euler HH, Marmont AM, Bacigalupo A et al. Early recurrence or persistence of autoimmune diseases after unmanipulated autologous stem cell transplantation. Blood 1996; 88(9):3621-5.

95. Burt RK, Slavin S, Burns WH et al. Induction of tolerance in autoimmune diseases by hematopoietic stem cell transplantation: getting closer to a cure? Blood 2002; 99(3):768-84.

96. De Stefano P, Zecca M, Giorgiani G et al. Resolution of immune haemolytic anaemia with allogeneic bone marrow transplantation after an unsuccessful autograft. Br J Haematol 1999; 106(4):1063-4.

97. Muller BU, Tichelli A, Passweg JR et al. Successful treatment of refractory acquired pure red cell aplasia (PRCA) by allogeneic bone marrow transplantation. Bone Marrow Transplant 1999; 23(11):1205-7.

98. Raetz E, Beatty PG, Adams RH. Treatment of severe Evans syndrome with an allogeneic cord blood transplant. Bone Marrow Transplant 1997; 20(5):427-9.

99. Oyama Y, Papadopoulos EB, Miranda M et al. Allogeneic stem cell transplantation for Evans syndrome. Bone Marrow Transplant 2001; 28(9):903-5.

100. Lee WY, Oh ES, Min CK et al. Changes in autoimmune thyroid disease following allogeneic bone marrow transplantation. Bone Marrow Transplant 2001; 28(1):63-6.

101. Shapiro E, Krivit W, Lockman L et al. Long-term effect of bone-marrow transplantation for childhood-onset cerebral X-linked adrenoleukodystrophy. Lancet 2000; 356(9231):713-8.

102. Bohl D, Bosch A, Cardona A et al. Improvement of erythropoiesis in beta-thalassemic mice by continuous erythropoietin delivery from muscle. Blood 2000; 95(9):2793-8.

103. Aiuti A, Slavin S, Aker M et al. Correction of ADA-SCID by stem cell gene therapy combined with nonmyeloablative conditioning. Science 2002; 296(5577):2410-3.

104. Krause DS, Theise ND, Collector MI et al. Multi-organ, multi-lineage engraftment by a single bone marrow-derived stem cell. Cell 2001; 105(3):369-77.

105. Glimm H, Oh IH, Eaves CJ. Human hematopoietic stem cells stimulated to proliferate in vitro lose engraftment potential during their S/G(2)/M transit and do not reenter G(0). Blood 2000; 96(13):4185-93.

106. Sadelain M, Frassoni F, Riviere I. Issues in the manufacture and transplantation of genetically modified hematopoietic stem cells. Curr Opin Hematol 2000; 7(6):364-77.

107. Brown MP, Topham DJ, Sangster MY et al. Thymic lymphoproliferative disease after successful correction of CD40 ligand deficiency by gene transfer in mice. Nat Med 1998; 4(11):1253-60.

108. Notarangelo LD, Giliani S, Mazza C et al. Of genes and phenotypes: the immunological and molecular spectrum of combined immune deficiency. Defects of the gamma(c)-JAK3 signaling pathway as a model. Immunol Rev 2000; 178:39-48. Review.

109. Buckley RH. Primary immunodeficiency diseases due to defects in lymphocytes. N Engl J Med 2000; 343(18):1313-24.

110. Moshous D, Callebaut I, de Chasseval R et al. Artemis, a novel DNA double-strand break repair/V(D)J recombination protein, is mutated in human severe combined immune deficiency. Cell 2001; 105(2):177-86.

111. Fischer A. Severe combined immunodeficiencies (SCID). Clin Exp Immunol 2000; 122(2):143-9.

112. Bertrand Y, Landais P, Friedrich W et al. Influence of severe combined immunodeficiency phenotype on the outcome of HLA nonidentical, T-cell-depleted bone marrow transplantation: a retrospective European survey from the European group for bone marrow transplantation and the european society for immunodeficiency. J Pediatr 1999; 134(6):740-8.

113. Haddad E, Le Deist F, Aucouturier P et al. Long-term chimerism and B-cell function after bone marrow transplantation in patients with severe combined immunodeficiency with B cells: A single-center study of 22 patients. Blood 1999; 94(8):2923-30.

114. Blaese RM, Culver KW, Miller AD et al. T lymphocyte-directed gene therapy for ADA- SCID: initial trial results after 4 years. Science 1995; 270(5235):475-80.

115. Kohn DB, Hershfield MS, Carbonaro D et al. T lymphocytes with a normal ADA gene accumulate after transplantation of transduced autologous umbilical cord blood CD34+ cells in ADA-deficient SCID neonates. Nat Med 1998; 4(7):775-80.

116. Hoogerbrugge PM, van Beusechem VW, Fischer A et al. Bone marrow gene transfer in three patients with adenosine deaminase deficiency. Gene Ther 1996; 3(2):179-83.

117. Bordignon C, Notarangelo LD, Nobili N et al. Gene therapy in peripheral blood lymphocytes and bone marrow for ADA- immunodeficient patients. Science 1995; 270(5235):470-5.

118. Di Santo JP. Inherited cytokine and cytokine receptor deficiencies in man. Int Rev Immunol 1998; 17(1-4):103-20.

119. Di Santo JP, Colucci F, Guy-Grand D. Natural killer and T cells of innate and adaptive immunity: lymphoid compartments with different requirements for common gamma chain- dependent cytokines. Immunol Rev 1998; 165:29-38. Review.

120. Di Santo JP, Rodewald HR. In vivo roles of receptor tyrosine kinases and cytokine receptors in early thymocyte development. Curr Opin Immunol 1998; 10(2):196-207.

121. Akashi K, Kondo M, von Freeden-Jeffry U et al. Bcl-2 rescues T lymphopoiesis in interleukin-7 receptor-deficient mice. Cell 1997; 89(7):1033-41.

122. Puel A, Leonard WJ. Mutations in the gene for the IL-7 receptor result in T(-)B(+)NK(+) severe combined immunodeficiency disease. Curr Opin Immunol 2000; 12(4):468-73.

123. Puel A, Ziegler SF, Buckley RH et al. Defective IL7R expression in T(-)B(+)NK(+) severe combined immunodeficiency. Nat Genet 1998; 20(4):394-7.

124. Mrozek E, Anderson P, Caligiuri MA. Role of interleukin-15 in the development of human CD56+ natural killer cells from CD34+ hematopoietic progenitor cells. Blood 1996; 87(7):2632-40.

125. Cavazzana-Calvo M, Hacein-Bey S, de Saint Basile G et al. Role of interleukin-2 (IL-2), IL-7, and IL-15 in natural killer cell differentiation from cord blood hematopoietic progenitor cells and from gamma c transduced severe combined immunodeficiency X1 bone marrow cells. Blood 1996; 88(10):3901-9.

126. Leonard WJ, Noguchi M, Russell SM et al. The molecular basis of X-linked severe combined immunodeficiency: The role of the interleukin-2 receptor gamma chain as a common gamma chain, gamma c. Immunol Rev 1994; 138:61-86. Review.

127. Soudais C, Shiho T, Sharara LI et al. Stable and functional lymphoid reconstitution of common cytokine receptor gamma chain deficient mice by retroviral-mediated gene transfer. Blood 2000; 95(10):3071-7.

128. Lo M, Bloom ML, Imada K et al. Restoration of lymphoid populations in a murine model of X-linked severe combined immunodeficiency by a gene-therapy approach. Blood 1999; 94(9):3027-36.

129. Otsu M, Anderson SM, Bodine DM et al. Lymphoid development and function in X-linked severe combined immunodeficiency mice after stem cell gene therapy. Mol Ther 2000; 1(2):145-53.

130. Hacein-Bey S, Basile GD, Lemerle J et al. gammac gene transfer in the presence of stem cell factor, FLT-3L, interleukin-7 (IL-7), IL-1, and IL-15 cytokines restores T-cell differentiation from gammac(-) X-linked severe combined immunodeficiency hematopoietic progenitor cells in murine fetal thymic organ cultures. Blood 1998; 92(11):4090-7.

131. Hacein-Bey H, Cavazzana-Calvo M, Le Deist F et al. gamma-c gene transfer into SCID X1 patients' B-cell lines restores normal high-affinity interleukin-2 receptor expression and function. Blood 1996; 87(8):3108-16.

132. Candotti F, Johnston JA, Puck JM et al. Retroviral-mediated gene correction for X-linked severe combined immunodeficiency. Blood 1996; 87(8):3097-102.

133. Taylor N, Uribe L, Smith S et al. Correction of interleukin-2 receptor function in X-SCID lymphoblastoid cells by retrovirally mediated transfer of the gamma-c gene. Blood 1996; 87(8):3103-7.

134. Hacein-Bey S, Gross F, Nusbaum P et al. Optimization of retroviral gene transfer protocol to maintain the lymphoid potential of progenitor cells. Hum Gene Ther 2001; 12(3):291-301.

135. Cavazzana-Calvo M, Hacein-Bey S, de Saint Basile G et al. Gene therapy of human severe combined immunodeficiency (SCID)-X1 disease. Science 2000; 288(5466):669-72.

136. Hacein-Bey-Abina S, Le Deist F, Carlier F et al. Sustained correction of X-linked severe combined immunodeficiency by ex vivo gene therapy. N Engl J Med 2002; 346(16):1185-93.

137. Mackall CL, Fleisher TA, Brown MR et al. Age, thymopoiesis, and CD4+ T-lymphocyte regeneration after intensive chemotherapy. N Engl J Med 1995; 332(3):143-9.

138. Schmidt M, Hoffmann G, Wissler M et al. Detection and direct genomic sequencing of multiple rare unknown flanking DNA in highly complex samples. Hum Gene Ther 2001; 12(7):743-9.

139. Notarangelo LD, Villa A, Schwarz K. RAG and RAG defects. Curr Opin Immunol 1999; 11(4):435-42.

140. Candotti F, Oakes SA, Johnston JA et al. In vitro correction of JAK3-deficient severe combined immunodeficiency by retroviral-mediated gene transduction. J Exp Med 1996; 183(6):2687-92.

141. Bunting KD, Lu T, Kelly PF et al. Self-selection by genetically modified committed lymphocyte precursors reverses the phenotype of JAK3-deficient mice without myeloablation. Hum Gene Ther 2000; 11(17):2353-64.

142. Yates F, Malassis-Seris M, Stockholm D et al. Gene therapy of RAG-2-/- mice: sustained correction of the immunodeficiency. Blood 2002; 100(12):3942-9.

143. D'Andrea AD, Grompe M. Molecular biology of Fanconi anemia: implications for diagnosis and therapy. Blood 1997; 90(5):1725-36.

144. Gluckman E, Auerbach AD, Horowitz MM et al. Bone marrow transplantation for Fanconi anemia. Blood 1995; 86(7):2856-62.

145. Guardiola P, Pasquini R, Dokal I et al. Outcome of 69 allogeneic stem cell transplantations for Fanconi anemia using HLA-matched unrelated donors: a study on behalf of the European Group for Blood and Marrow Transplantation. Blood 2000; 95(2):422-9.

146. Liu JM, Kim S, Read EJ et al. Engraftment of hematopoietic progenitor cells transduced with the Fanconi anemia group C gene (FANCC). Hum Gene Ther 1999; 10(14):2337-46.

147. Gregory Jr JJ, Wagner JE Verlander PC et al. Somatic mosaicism in Fanconi anemia: evidence of genotypic reversion in lymphohematopoietic stem cells. Proc Natl Acad Sci USA 2001; 98(5):2532-7.

148. Liu JM, Young NS, Walsh CE et al. Retroviral mediated gene transfer of the Fanconi anemia complementation group C gene to hematopoietic progenitors of group C patients. Hum Gene Ther 1997; 8(14):1715-30.

149. Gush KA, Fu KL, Grompe M et al. Phenotypic correction of Fanconi anemia group C knockout mice. Blood 2000; 95(2):700-4.

150. Segal BH, Leto TL, Gallin JI et al. Genetic, biochemical, and clinical features of chronic granulomatous disease. Medicine (Baltimore) 2000; 79(3):170-200.

151. Rio P, Segovia JC, Hanenberg H et al. In vitro phenotypic correction of hematopoietic progenitors from Fanconi anemia group A knockout mice. Blood 2002; 100(6):2032-9.

152. Winkelstein JA, Marino MC, Johnston Jr RB et al. Chronic granulomatous disease Report on a national registry of 368 patients. Medicine (Baltimore) 2000; 79(3):155-69.

153. Horwitz ME, Barrett AJ, Brown MR et al. Treatment of chronic granulomatous disease with nonmyeloablative conditioning and a T-cell-depleted hematopoietic allograft. N Engl J Med 2001; 344(12):881-8.

154. Ozsahin H, von Planta M, Muller I et al. Successful treatment of invasive aspergillosis in chronic granulomatous disease by bone marrow transplantation, granulocyte colony-stimulating factor-mobilized granulocytes, and liposomal amphotericin-B. Blood 1998; 92(8):2719-24.

155. Buescher ES, Alling DW, Gallin JI. Use of an X-linked human neutrophil marker to estimate timing of lyonization and size of the dividing stem cell pool. J Clin Invest 1985; 76(4):1581-4.

156. Bjorgvinsdottir H, Ding C, Pech N et al. Retroviral-mediated gene transfer of gp91phox into bone marrow cells rescues defect in host defense against Aspergillus fumigatus in murine X-linked chronic granulomatous disease. Blood 1997; 89(1):41-8.

157. Dinauer MC, Li LL, Bjorgvinsdottir H et al. Long-term correction of phagocyte NADPH oxidase activity by retroviral-mediated gene transfer in murine X-linked chronic granulomatous disease. Blood 1999; 94(3):914-22.

158. Mardiney 3rd M, Jackson SH, Spratt SK et al. Enhanced host defense after gene transfer in the murine p47phox-deficient model of chronic granulomatous disease. Blood 1997; 89(7):2268-75.

159. Malech HL, Maples PB, Whiting-Theobald N et al. Prolonged production of NADPH oxidase-corrected granulocytes after gene therapy of chronic granulomatous disease. Proc Natl Acad Sci USA 1997; 94(22):12133-8.

160. Kelly PF, Vandergriff J, Nathwani A et al. Highly efficient gene transfer into cord blood nonobese diabetic/severe combined immunodeficiency repopulating cells by oncoretroviral vector particles pseudotyped with the feline endogenous retrovirus (RD114) envelope protein. Blood 2000; 96(4):1206-14.

161. Kiem HP, Heyward S, Winkler A et al. Gene transfer into marrow repopulating cells: comparison between amphotropic and gibbon ape leukemia virus pseudotyped retroviral vectors in a competitive repopulation assay in baboons. Blood 1997; 90(11):4638-45.

162. Miyoshi H, Smith KA, Mosier DE et al. Transduction of human CD34+ cells that mediate long-term engraftment of NOD/SCID mice by HIV vectors. Science 1999; 283(5402):682-6.

163. Saulnier SO, Steinhoff D, Dinauer MC et al. Lentivirus-mediated gene transfer of gp91phox corrects chronic granulomatous disease (CGD) phenotype in human X-CGD cells. J Gene Med 2000; 2(5):317-25.
164. Abonour R, Williams DA, Einhorn L et al. Efficient retrovirus-mediated transfer of the multidrug resistance 1 gene into autologous human long-term repopulating hematopoietic stem cells. Nat Med 2000; 6(6):652-8.
165. Jin L, Siritanaratkul N, Emery DW et al. Targeted expansion of genetically modified bone marrow cells. Proc Natl Acad Sci USA 1998; 95(14):8093-7.
166. Leboulch P, Huang GM, Humphries RK et al. Mutagenesis of retroviral vectors transducing human beta-globin gene and beta-globin locus control region derivatives results in stable transmission of an active transcriptional structure. EMBO J 1994; 13(13):3065-76.
167. Sadelain M, Wang CH, Antoniou M et al. Generation of a high-titer retroviral vector capable of expressing high levels of the human beta-globin gene. Proc Natl Acad Sci USA 1995; 92(15):6728-32.
168. Rivella S, Sadelain M. Genetic treatment of severe hemoglobinopathies: the combat against transgene variegation and transgene silencing. Semin Hematol 1998; 35(2):112-25.
169. Kalberer CP, Pawliuk R, Imren S et al. Preselection of retrovirally transduced bone marrow avoids subsequent stem cell gene silencing and age-dependent extinction of expression of human beta-globin in engrafted mice. Proc Natl Acad Sci USA 2000; 97(10):5411-5.
170. May C, Rivella S, Callegari J et al. Therapeutic haemoglobin synthesis in beta-thalassaemic mice expressing lentivirus-encoded human beta-globin. Nature 2000; 406(6791):82-6.
171. Samakoglu S, Fattori E, Lamartina S et al. betaMinor-globin messenger RNA accumulation in reticulocytes governs improved erythropoiesis in beta thalassemic mice after erythropoietin complementary DNA electrotransfer in muscles. Blood 2001; 97(8):2213-20.
172. May C, Rivella S, Chadburn A et al. Successful treatment of murine beta-thalassemia intermedia by transfer of the human beta-globin gene. Blood 2002; 99(6):1902-8.
173. Imren S, Payen E, Westerman KA et al. Permanent and panerythroid correction of murine beta thalassemia by multiple lentiviral integration in hematopoietic stem cells. Proc Natl Acad Sci USA 2002; 99(22):14380-5.
174. Pawliuk R, Westerman KA, Fabry ME et al. Correction of sickle cell disease in transgenic mouse models by gene therapy. Science 2001; 294(5550):2368-71.
175. Brazelton TR, Rossi FM, Keshet GI et al. From marrow to brain: expression of neuronal phenotypes in adult mice. Science 2000; 290(5497):1775-9.
176. Ferrari G, Cusella-De Angelis G, Coletta M et al. Muscle regeneration by bone marrow-derived myogenic progenitors. Science 1998; 279(5356):1528-30.
177. Galli R, Borello U, Gritti A et al. Skeletal myogenic potential of human and mouse neural stem cells. Nat Neurosci 2000; 3(10):986-91.
178. Gussoni E, Soneoka Y, Strickland CD et al. Dystrophin expression in the mdx mouse restored by stem cell transplantation. Nature 1999; 401(6751):390-4.
179. Mezey E, Chandross KJ, Harta G et al. Turning blood into brain: cells bearing neuronal antigens generated in vivo from bone marrow. Science 2000; 290(5497):1779-82.
180. Orlic D, Kajstura J, Chimenti S et al. Mobilized bone marrow cells repair the infarcted heart, improving function and survival. Proc Natl Acad Sci USA 2001; 98(18):10344-9.
181. Lagasse E, Connors H, Al-Dhalimy M et al. Purified hematopoietic stem cells can differentiate into hepatocytes in vivo. Nat Med 2000; 6(11):1229-34.

Index